Rubbish Be

'A compelling ethnography of Uruguayan waste pickers. This important intervention asks who has the economic and moral right to the surplus and excess that drive capitalism. As O'Hare shows, the waste pickers lay claim to this resource as part of a dialogue with environmental and social justice, through practices of care and communing.'

—Catherine Alexander, Department of Anthropology, University of Durham

'Written with a clear and convincing prose, this book makes a major contribution to and advances waste studies, environmental studies, and the anthropology of infrastructure by updating our extant theories of labour, the economy and the commons. It will not only serve as a useful teaching resource but also as a model for future scholars.'

—Zsuzsa Gille, Professor of Sociology and Director of Global Studies, University of Illinois at Urbana-Champaign

'Activist scholarship of the highest calibre. This is an intimate, humorous and razor-sharp analysis of the politics of urban waste. O'Hare mounts a passionate defence of waste as commons, in the face of corporate and state initiatives to reconfigure waste as resource.'

—Penny Harvey, Professor of Social Anthropology, University of Manchester

'By lingering with waste workers in Montevideo, Uruguay, O'Hare intricately unfolds the changing conditions of rubbish as it circulates through scavenging practices, urban infrastructures, circular economies, and global property structures. *Rubbish Belongs to the Poor* offers a radically different view of how to shape environmental citizenships.'

—Jennifer Gabrys, Chair in Media, Culture and Environment, University of Cambridge

'Radically rethinks the commons, urban infrastructure and waste in ways that hold significant political implications for our time. Patrick O'Hare calls us to take seriously the work of waste reclaimers not as a problem in need of a solution, but rather, as a source of a new kind of politics.'

—Kathleen Millar, Department of Sociology and Anthropology, Simon Fraser University

Anthropology, Culture and Society

Series Editors:
Jamie Cross, University of Edinburgh,
Holly High, Deakin University
and
Joshua O. Reno, Binghamton University

Recent titles:

The Limits to Citizen Power: Participatory Democracy and the Entanglements of the State
Victor Albert

Vicious Games: Capitalism and Gambling
Rebecca Cassidy

Anthropologies of Value
Edited by Luis Fernando Angosto-Ferrandez and Geir Henning Presterudstuen

Ethnicity and Nationalism: Anthropological Perspectives
Third Edition
Thomas Hylland Eriksen

Small Places, Large Issues: An Introduction to Social and Cultural Anthropology
Fourth Edition
Thomas Hylland Eriksen

What is Anthropology?
Second Edition
Thomas Hylland Eriksen

Deepening Divides: How Territorial Borders and Social Boundaries Delineate Our World
Edited by Didier Fassin

At the Heart of the State: The Moral World of Institutions
Didier Fassin, et al.

Anthropology and Development: Challenges for the Twenty-first Century
Katy Gardner and David Lewis

Children of the Welfare State: Civilising Practices in Schools, Childcare and Families
Laura Gilliam and Eva Gulløv

Faith and Charity: Religion and Humanitarian Assistance in West Africa
Edited by Marie Nathalie LeBlanc and Louis Audet Gosselin

Private Oceans: The Enclosure and Marketisation of the Seas
Fiona McCormack

Grassroots Economies: Living with Austerity in Southern Europe
Edited by Susana Narotzky

The Rise of Nerd Politics: Digital Activism and Political Change
John Postill

Base Encounters: The US Armed Forces in South Korea
Elisabeth Schober

Ground Down by Growth: Tribe, Caste, Class and Inequality in Twenty-First-Century India
Alpa Shah, Jens Lerche, et al

Watershed Politics and Climate Change in Peru
Astrid B. Stensrud

When Protest Becomes Crime: Politics and Law in Liberal Democracies
Carolijn Terwindt

Race and Ethnicity in Latin America
Second Edition
Peter Wade

Rubbish Belongs to the Poor

Hygienic Enclosure and the Waste Commons

Patrick O'Hare

First published 2022 by Pluto Press
New Wing, Somerset House, Strand, London WC2R 1LA

www.plutobooks.com

British Library Cataloguing in Publication Data
A catalogue record for this book is available from the British Library

ISBN 9780745341408 Paperback
ISBN 9780745341385 Hardback
ISBN 9781786807489 PDF eBook
ISBN 9781786807496 EPUB eBook

Typeset by Stanford DTP Services, Northampton, England

Simultaneously printed in the United Kingdom and United States of America

Contents

Figures

Series Preface

As people around the world confront the inequality and injustice of new forms of oppression, as well as the impacts of human life on planetary ecosystems, this book series asks what anthropology can contribute to the crises and challenges of the twenty-first century. Our goal is to establish a distinctive anthropological contribution to debates and discussions that are often dominated by politics and economics. What is sorely lacking, and what anthropological methods can provide, is an appreciation of the human condition.

We publish works that draw inspiration from traditions of ethnographic research and anthropological analysis to address power and social change while keeping the struggles and stories of human beings centre stage. We welcome books that set out to make anthropology matter, bringing classic anthropological concerns with exchange, difference, belief, kinship and the material world into engagement with contemporary environmental change, capitalist economy and forms of inequality. We publish work from all traditions of anthropology, combining theoretical debate with empirical evidence to demonstrate the unique contribution anthropology can make to understanding the contemporary world.

Jamie Cross, Holly High and Joshua O. Reno

Acknowledgements

This book was made possible by the openness, trust, and friendship of the clasificadores of Felipe Cardoso and its surrounding neighbourhoods of Flor de Maroñas and the Cruz de Carrasco. To Marcos Borges and Alejandra Martínez in particular: thank you for being such wonderful neighbours, and making us feel so at ease. I am full of admiration for your generosity, hard work, and good humour as you raise your beautiful family. There are many other families in COVIFU, and regular visitors to the cooperative such as the Méndez family and cousin 'Checo', who also opened their doors to us, have my utmost respect, and made of Montevideo a place that will always feel like home. Thank you to the clasificadores at the cantera, at COFECA, at the 'Planta Aries', and beyond who allowed me to work alongside them even though I couldn't carry a *bolsón* to save my life. I am also grateful to the tireless campaigners of the UCRUS who allowed me to join their meetings at the Galpón de Corrales, their marches and their neighbourhood sorties. My thanks also go to the staff at the Intendencia de Montevideo, the Ministerio de Desarrollo Social (MIDES), and all the other institutions and NGOs such as Juventud Para Christo who allowed me to conduct fieldwork inside and alongside their organisations.

It will of course not be possible to name everyone who assisted with the research out of which this book grew but let me try to name a few of them: the extended Ponce de León family, the Almada family, the Umpiérrez family, the lovely Santos clan, the Pérez family, the Mora family, the Juvencio family, Veronica Rodríguez, Jorge Meoni, Nacho, Gabriel and the rest of the team at Los Treboles, Raúl and Ana, Moncho and Jennifer, Jorge and Luján, Paula and Tungi, Julian and Laura, Iris, the Hermanas Franciscanas who live and do great work in the Cruz de Carrasco, Nito, Esteban and the boys from Joaquín de la Sagra, Selva, Uncle Luca, Fabi, and the other neighbours from the Felipe Cardoso asentamiento, and all the housing cooperative kids who filled our house with paint, laughter, and fun. Thanks to Ramita, Coco, and Noño for letting is hang out with them at the feria, and to older clasificadores such as Claudia, Carlos, Lucía, Coco, China, Zuli and others for sharing their memories of a life classifying and recovering discards.

At the University of Cambridge, the doctoral thesis from which this book grew benefitted from discussions with a range of colleagues. Sian Lazar was an excellent doctoral supervisor who was always supportive of my initiatives and provided invaluable commentary and input, without which this book would not be what it is. Further thanks go to Oliver, Felix, Sofía, Laurie, Matt, and Eduardo for their comments on chapter drafts at different points, and to Clara, Teo, John, Corinna, Patrick, Anthony, Chloe and Christos for proofreading. Thanks to Christos for your intellectual support and inspiration, to my study buddy and little sister Fiona, and to my Argentine counterpart Santiago for friendship and collaboration. A short stay at the Nucleo de Pesquisas em Cultura e Economia (NuCEC) in Rio de Janeiro was invaluable for the critical commentary received. Clare College, the Centre of Latin American studies (CLAS), and the Division of Social Anthropology provided additional financial support during my doctoral studies and their administrative staff were also incredibly friendly and helpful during that time.

Over the years, my research has been generously financed by the Economic and Social Research Council (ESRC) through a 1+3 masters and doctoral studentship and a subsequent postdoctoral fellowship (PDF), which enabled me to work on the proposal for this book and begin to transform the doctoral thesis into a book manuscript. The latter was hosted by the Department of Social Anthropology at the University of Manchester, where I encountered a thriving intellectual community and benefited from critical feedback on chapter drafts from colleagues and my institutional mentor Penelope Harvey. Penny was also the external examiner for my doctoral thesis and together with my internal examiner Andrew Sanchez, convinced me that the thesis should see the light of day as a monograph. The one-year PDF allowed me to translate one of the chapters of this book into Spanish and present it to an audience of research participants in Montevideo. That presentation was facilitated by Lucia Férnandez and Gerardo Sarachu, whose work on and with Uruguayan wastepickers remains inspirational and who have always been incredibly generous with their time and friendship.

My colleagues on the two research projects that I have formed part of as this book progressed – Lucy and Alex from the Cartonera Publishing project and Brigitte and Teresa from Cambridge Circular Plastics – are also due thanks. Even if they did not directly comment on this manuscript, their friendship, collegiality, and belief gave me the confidence to take this book project forward. From an early stage, this book also had the confidence, support and input of Josh Reno, series editor at Pluto Press and an inspiration in the anthropology of waste. As the proposal progressed, many more people from Pluto played pivotal roles in bringing this project to fruition, including David Castle, Robert Webb, and Melanie Patrick. The book also benefitted from the valuable suggestions and comments of three anonymous reviewers and the detailed feedback of Catherine Alexander, who waived her anonymity in order to facilitate engaging discussions that have gone beyond this book and resulted in rewarding collaborations elsewhere.

My current UKRI Future Leaders Fellowship (FLF) enabled me to put the finishing touches to this manuscript whilst embarking on a new project on recycled plastics and the circular economy. The fellowship has also brought me full circle back to the University of St Andrews, where I studied for my undergraduate degree, and whose exchange programme took me to Uruguay where I began to work with clasificadores for the first time. The response to the oft-posed question 'why Uruguay?' thus has to include Professor Gustavo San Román, who organised the exchange program with the Universidad de la República in Montevideo. My return to St Andrews, and indeed my career in social anthropology, has been made possible by the continued support and mentorship of Professor Mark Harris. I also thank my old dumpster diver pals from St Andrews whose memories I jogged when preparing this manuscript.

My utmost thanks go out to Mary, who I met at my first academic event in Cambridge. What began as flirtation in social science methods classes soon grew into a gran amor. Although she is not overly present in this book, she was a wonderfully warm presence in Montevideo, where we built our first home together in the shadow of Felipe Cardoso. 'Fancy coming out and living next to a dump for a year' is not a question usually answered in the affirmative, but Mary has never been shy of adventure, and fully embraced what was in many ways an idyllic life in the housing cooperative. Several neighbours said that it had taken Mary to turn a house into a home and I shudder to think what the house or fieldwork would have been like without her. I probably still wouldn't have gotten round to buying a toilet seat! Needless to say, Mary has also provided the most consistent and important opinions as this book has taken shape.

Mary's family probably didn't know what they had gotten themselves in for when I first came around for tea, as within a few months I was moving in. My heartfelt thanks go out to Françoise, who opened up her home for me in Cambridge and allowed me make it my own, as well as providing valuable feedback on this manuscript at various moments. Help, love, and support from my own brothers and parents have also never been in short supply. Mary and I have created our own little family with the arrival of our two daughters, Rosie and Alba. They are a constant a source of light and joy. I only hope that this book does justice to the support and love of this wonderful group of people that I have been fortunate enough to have around me over recent years.

Abbreviations

AUIP	Asociación Uruguaya de Industrias del Plástico (Uruguayan Plastics Industry Association)
CAP	Consorcio Ambiental del Plata (Plate Environmental Consortium)
CCT	compulsory competitive tendering
CEGRU	Cámara Empresarial de Gestores de Residuos (Chamber of Uruguayan Waste Managers)
CIU	Cámara de Industrias del Uruguay (Uruguayan Chamber of Commerce)
COFECA	Cooperativa Felipe Cardoso (Felipe Cardoso Cooperative)
COVIFU	Cooperativa de Vivienda Nuestro Futuro (Our Future Housing Cooperative)
CPT	Cristo Para Todos (Christ for All)
EPA	Environmental Protection Act
EU	European Union
FORU	Federación Obrera Regional Uruguaya (Uruguayan Regional Workers Federation)
GTC	Grupo de Trabajo con Clasificadores (Clasificador Working Group)
HDPE	high-density polyethylene
IDB	Inter-American Development Bank
IMM/IM	Intendencia Municipal de Montevideo (later Intendencia de Montevideo)
INEFOP	Instituto Nacional de Empleo y Formación Profesional (National Institute for Employment and Professional Development)
LAWDCs	local authority waste disposal companies
MIDES	Ministerio de Desarrollo Social (Ministry of Social Development)
MNCR	Movimento Nacional dos Catadores de Materiais Recicláveis (National Movement of Recyclable Material Classifiers – Brazil)
MRF	materials recovery facility

MSW	municipal solid waste
MTE	Movimiento de Trabajadores Excluidos (Excluded Workers Movement – Argentina)
NGO	non-governmental organisation
OPP	Oficina de Planeamiento y Presupuesto (Planning and Budgeting Office)
PET	polyethylene terephthalate
PIT-CNT	Plenario Intersindical de Trabajadores – Convención Nacional de Trabajadores (National Trade Union Federation – Uruguay)
PNUD	Programa de Naciones Unidas de Desarrollo (United Nations Development Program)
PVC	polyvinyl chloride
Red-LACRE	Red Latinoaméricano y del Caribe de Recicladores (Latin American and Caribbean Recyclers Network)
SDF	Sitio de Disposición Final (Final Disposal Site)
UCRUS	Unión de Recicladores de Residuos Urbanos Solidos (Union of Urban Solid Waste Recyclers – Uruguay)

Introduction: 'La Basura es de los Pobres!' – *'Rubbish Belongs to the Poor'*

'*Rubbish belongs to the poor!*' I first heard this slogan filling the air with its clamour when, in 2014, I attended one of several road blocks organised in Montevideo by the Union of Urban Solid Waste Classifiers (Unión de Clasificadores de Residuos Urbanos Sólidos – UCRUS),[1] the trade union for Uruguay's waste-pickers. A junction at the entrance to the neighbourhood of El Cerro was blocked off by a dozen or so waste-picking families, their horses and their carts. El Cerro, named after the hill (*cerro*) adjacent to which it was built, is an iconic neighbourhood. The very name of Uruguay's capital, Montevideo, supposedly originates from the sighting of the hill (*monte*) by approaching Portuguese sailors in the sixteenth century, and the military fort built there features on Uruguay's national coat of arms. The neighbourhood was founded in the nineteenth century as Cosmópolis, a utopian settlement built to house worker-immigrants from far-flung lands, and the names of its narrow streets still recall this cosmo-proletarian dream: Russia, Egypt, Greece, England, Cuba. During the nineteenth, and early twentieth century, El Cerro became synonymous with the city's meat-packing industry and a stronghold of worker militancy and anarcho-syndicalism. Nowadays, it is home to a new recycling plant, and a large number of waste-pickers, known in Uruguay as *clasificadores* (classifiers), many of whom had gathered for the roadblock (Figure 1).

The claim that 'rubbish belongs to the poor!' was shouted insistently by a group of children sitting astride a horse and cart, a vehicle that their family normally used to collect recyclables in the city. The children's father told me that new 'hermetically sealed' rubbish containers that were being rolled out in the city centre by the local government or *Intendencia* – and into which citizens had to separate their recyclable and non-recyclable waste – were making life difficult for waste-pickers like his family, and purposefully so. Although the containers were being presented

Figure 1 The roadblock in the El Cerro neighbourhood
Source: Author photo, 15 August 2014.

by municipal authorities as 'anti-vandal', a representative told me that they had been specially altered by an Italian engineering firm in order to stop children and adult waste-pickers from entering. Unfortunately, the narrow-mouthed design also meant that it was equally impossible to deposit a bag of recyclables inside.

They were installed by the private company that operates the concession for the collection and disposal of waste in central Montevideo, along with accompanying trucks that came with a price tag of over US$500,000 each. Materials from the new containers were being channelled to the clasificadores employed in newly inaugurated formal sector plants but spaces at the plants were limited, and whole neighbourhoods were being closed off to informal sector kerbside waste-pickers, the father complained. Commercial enterprises were already forbidden from giving informal sector actors waste by a 2012 municipal decree, which stipulated the need to contract a formalised waste transport company. What livelihood were they expected to turn to instead, his wife asked: theft or drug-dealing?

In this series of short statements, the family from El Cerro touched on the central issues that have brought Montevidean waste-pickers onto the streets to protest in recent years: problematic new containers, restrictions on the circulation of horses and carts, and the diversion of recyclables to formalised waste-pickers in recycling plants. They also hinted at the moral economy of waste-picking labour, which was defended by union leaders as honourable work and contrasted with drug-dealing and theft, cited as the only feasible alternatives available to those from Montevideo's shantytowns. The speakers and the context of the 'rubbish belongs to the poor' claim – children astride a horse – are important, for these are human and non-human subjects whose engagement in urban waste-work has been singled out as unacceptable in twenty-first-century Uruguay, clashing with what I go on to describe as Montevideo's 'infrastructural modernity'. Nor is their use of the word 'rubbish' inconsequential. In the following pages, I argue that the peculiar and problematic status of rubbish cannot simply be replaced by the increasing, enthusiastic move to reconceptualise waste as resource. The word 'poor' (*pobres*), a descriptive term long discarded by Uruguayan policy makers in favour of 'marginalised' and 'excluded' was in turn still an important self-referential term used by Montevideo's popular sectors, and one that I will argue characterised them as customary beneficiaries of access to a waste commons.

Many clasificadores recover recyclables using a horse and cart, something which first attracted my attention when I was a young undergraduate on a student exchange year in Uruguay in 2009/10. My interest piqued, I attended the screening of a film about clasificadores and was subsequently invited to work alongside the Cooperativa Pedro Trastos, a small, family-based waste-picking cooperative that operated from a base near the landfill in the Cruz de Carrasco neighbourhood. This was my first entry point into Uruguayan waste-picking: a world of workshops about the Rochdale pioneers, of painstaking recovery of items as small as receipts, of heavy bags laden with recyclables hoisted onto shoulders, of supplies of shampoo recovered from the discards of Montevideo's Unilever plant. My summer with 'Los Trastos'[2] led me to establish strong foundations and friendships in and around the landfill as I doubled up as a labourer in the construction of the waste-pickers' cooperative homes.

When I returned for my doctoral research in 2014, I was fondly remembered – alas not for my skills in the building trade. When it came to waste-picking however, I already had a bit of experience. As an undergraduate student and activist at the University of St Andrews, I had regularly 'dumpster dived' into the bountiful bins of local supermarkets. Living in shared student accommodation and attending regular £1 vegetarian lunches organised by environmental activists, a surprising amount of our ingredients were rummaged rather than purchased. We learned to use our senses (smell, taste, touch) to pick out what still seemed edible, to embrace certain food categories as relatively safe (fruit, vegetables) and cast a more discerning eye over others (meat, dairy). If the poor made a historic claim to ownership over Montevideo's waste, food waste in St Andrews at least partially belonged to skint students. Certainly, the needs of fellow students subsisting on government loans were different from those of my Uruguayan friends but, with both groups of fellow-travellers, I encountered disbelief at the colossal amounts of stuff being thrown away, common sensory practices of recovery, and resistance on the part of authorities to our attempts to channel materials away from landfill.

In St Andrews, our prime site for recovery was the bins of a certain high street supermarket. Given the CCTV cameras in operation, face coverings were essential, long before they became part of the everyday shopping experience in times of COVID-19. Anarchist friends enjoyed pulling on hoods and balaclavas in the black of night, using torches to assess the contents of bins and then packing them into backpacks to be examined in more detail at the kitchen table at home. The supermarket managers, nonplussed at our activity, decided to lock the bins shut. A friend glued the padlocks in turn, leading to the removal of the locks and our triumphant return to the rummage. Students 1, supermarket 0.

It would be all too easy to dismiss the comparison between the activities carried out by Uruguayan waste-pickers and St Andrews activists as fundamentally different types of action, the first driven by desperate need, the second by a combination of political activism and temporary, extremely relative, hardship. This book chooses a different path, recognising common processes of enclosure in the Global North and South that seek to restrain our access, and particularly that of vulnerable groups, to the excesses of production. Although it draws on in-depth ethnography, its subject matter is a global system that seeks to protect the private nature of property even as it is carted off to landfill, and creates opportunities for profiteering in waste disposal rather than engineering an equitable distribution of surplus. It also asks how we might perceive the activity of scavenging – so long considered an affront to human dignity – in a radically different way, thus transforming what Jacques Rancière (2004) calls the 'distribution of the sensible' (2004) with regard to waste. Might Uruguayan clasificadores be considered, in the words of the priest who has worked most closely with them, 'ecological prophets' (Alonso 1992)? By bringing together the common pleasures found in rummaging and rescuing things that have lost their commodity value but not their identities, we explore how new forms of responsible environmental citizenship might be forged.

Getting our hands dirty

The small South American country of Uruguay, with a population of just under 3.5 million people, may not be particularly familiar to readers. If you are anything like my Glaswegian grandfather, the name Uruguay might conjure up any one of three things: Fray Bentos corned beef, the Second World War German battleship *Admiral Graf Spee*, which was scuttled in Montevideo bay, or football, with Uruguay hosting and winning the first ever World Cup in 1930, followed up by a famous second victory in Brazil's Maracanã stadium in 1950. Uruguayans are a lot more familiar with Britain at least, since in some accounts the country owes its existence to British efforts to create a buffer state between regional powerhouses Argentina and Brazil. After the 500 day-long Cisplatine war between Uruguayan independentists and present-day Argentines on one side, and Brazil on the other, it was Britain and its Viscount Ponsonby who mediated the 1825 Treaty of Montevideo, which gave the country its independence. The links that I seek to make in this book between Uruguay and the UK are thus not as unexpected as they might appear. And just as in world football Uruguay continually punches above its weight, I seek to argue here that humble waste-pickers, clasificadores, might help us to redefine ideas of social justice, dispossession, and dignified life in the twenty-first century, far beyond the boundaries of what is affectionately known as *el paísito*: the little country.

For this, I'm afraid we're going to have to get our hands dirty. So let us dive directly into Uruguay's rubbish and the labour of its classifiers with a typical afternoon at the Cooperativa Felipe Cardoso (COFECA). This is a waste-picker cooperative situated adjacent to Montevideo's Felipe Cardoso landfill. It was March of 2014, and I worked assiduously, emptying out black plastic bags delivered by municipal trucks onto ground moist from days of rain. Something fell in front of me, and I attempted to classify it visually. A balloon? No. A condom? No, these were more likely to be found alongside little bars of soap and damp hand-towels in the bags that the cooperative received from love hotels. On closer inspection, the clear tubing identified this item as a medical discard. Composite plastics made it unrecyclable, so I left it in peace, untying another black bag instead. I was working alongside Pedro, affectionately nicknamed Grampa (El Abuelo) by his colleagues. Grampa had returned to waste-picking at 60 after employers had swindled him out of a pension. With skilful dexterity, he unpicked one of the white sacks that arrived from industrial bakeries.

Such sacks are usually fairly promising and might contain a family's month's supply of flour. Yet they could also hold loose dough that stuck uncomfortably to gloved fingers. Many of the black plastic bags, meanwhile, held an assortment of items factory workers had placed into small waste-paper baskets. Often, these were simply scrunched-up pieces of paper used to wipe a surface, a nose, or perhaps a bottom. They were effectively worthless, the lowest quality types of paper made by IPUSA, Uruguay's largest tissue manufacturer, where some of the white paper that COFECA classified was sent to be recycled. A transparent plastic packet carried the logo of one of IPUSA's brands: Elite tissues. I left empty crisp packets and tobacco pouches on the ground but picked up clear plastic PET (polyethylene terephthalate) bottles, putting them into one of the large plastic bags that I had placed around me, in a category clasificadores call 'the little bottle' (*la botellita*). The black plastic bag itself joined others in a category known as 'black nylon' (*nailon negro*) and I pulled at another that had become snagged on the adhesive from a disposable nappy. From a distance, I heard my colleague Matute singing in an attempt to liven up the mood of a landscape that appeared particularly grim after a deluge: mud mixing with soggy cardboard and moist food waste. Still, workers had a solid concrete floor under foot that they had proudly laid themselves several years previous. They also had protective 'raincoats' on: a mixture of cagouls and black bin bags in which they had fashioned holes for heads and arms, giving the impression that the rubbish was engaged in the labour of its own classification.

I pushed a mass of transparent 'white nylon' (*nailon blanco*) into a different bag – this has a higher market value as long as it isn't too dirty. Another black bag was mixed with the most common and unpleasant contaminant one is likely to find in Uruguayan household rubbish, mundane rather than toxic: the damp, scattered leaves from *yerba mate* tea.[3] I speculated on the origin of this bag: workers, sitting around a table enjoying pre-prepared sandwiches and quiches on polystyrene trays that they disposed of in the bin, joined soon after by plastic cups from an office cooler, then by the post-lunch *mate* leaves and a tea bag or two? As the French philosopher François Dagognet (1997: 13) asserts, 'even the smallest utensil, like the most used cloth, carry with them a sort of tattoo indicating time and contact' and in such conditions 'the abandoned or the now unemployed seem an incomparable witness'. Waste here appears as an indicative, if ultimately unreliable, archive (Rathje and Murphy 2001: 4, Yaeger 2003).

Stevie, my neighbour in the nearby Cooperativa de Vivienda Nuestro Futuro (COVIFU) housing cooperative as well as my co-worker at COFECA, offered me a bag smeared with a white cream and enclosing an unidentifiable meaty substance inside. 'For my dog?', I asked. 'No, the chickens', he suggested, throwing it towards a pile of potential take-home things known as *requeche* that I followed other workers in making for myself. This term is Uruguayan slang for 'leftovers'. As used by the general population, it most commonly refers to food. Among clasificadores, however, *requeche* can refer to anything recovered from the trash that can be consumed or has a re-use value. A horse wandered over to pick at the fringes of organic waste, and Stevie gently shooed him away by making a puckering noise with his lips. I tried to engage my workmate in conversation but he proved reluctant, preferring instead to communicate with the horse and wipe the cream from his hands. He stopped, and with a colleague began to tie up a bundle of cardboard that he had piled on top of a canvas. I continued unenthusiastically picking at the damp pile in front of me. It was an unsatisfying load, because the small bags were too fiddly to open, I knew that I would not find much of value inside, and whatever I did find had been soiled by the ubiquitous *mate* leaves. I stopped to help the others hoist the cardboard canvas onto Stevie's shoulders.

The initial lift is a four-person job but the carrying was done by Stevie alone. The white cream had got onto the canvas and smudged his neck and shoulder. The bag was heavy because the cardboard had become wet, but this was good because workers are paid by the kilo, and so will earn more from sodden cellulose. Raindrops started to fall again and suddenly there were dark clouds rising over the old landfill of Usina 6, now a grassy hill whose raison d'être is suggested only by the methane pipes that emerge from deep inside its belly. '*Se viene el aguaaaaa*' ('The water is comiiiiiing!'), Matute shouted, interrupting his broadcast of cumbia hits with a weather update. The downpour was not heavy enough to justify stopping work, so we continued, emptying black bag after black bag. Stevie laid down another canvas onto which I started to pile more damp cardboard, which he arranged so as to maximise the load that could be carried. 'Fragile', read the letters on the cardboard, a description that also applied to Stevie's body as he bore the heavy load, although he was unlikely to admit it.

Whether sneaking into the landfill, joining *compañeros* in cooperatives, manning the conveyor belt at recycling plants, levelling piles of rubbish at family recycling yards, or even making the odd outing on horse and cart, I have come to know Montevideo's waste-stream rather more intimately than most of its residents. Waste-picking wasn't particularly easy and neither was explaining to a range of interlocutors exactly why I was so keen to get my hands dirty in Uruguayan rubbish. 'Is there no good *requeche* in Scotland?', a kindly waste-picker known as Uncle Luca laughed, as we greedily shared some broken Easter egg chocolate that had made its way to COFECA. Certainly, Easter was a good time of year, one in which we blessed the thin fragility of Easter eggs. But labour was, alongside friendship, also the most obvious 'thing' that I could offer waste-pickers who agreed to participate in my research, whether through recorded interviews, casual conversations, or simply tolerating my presence. There is a practical and an ethical component to such involvement too: I don't like standing idly by or getting in the way while others are working, even if observation is a key part of the anthropological endeavour. I wanted to pull my weight, even if I struggled and stumbled with heavy bags of recyclables on my shoulders, adopting what Walter Benjamin (1999: 364) called the 'jerky gait' of the ragpicker.

The base for my exploration was a home at the COVIFU, the 'Our Future Housing Cooperative', situated a few hundred metres from Montevideo's Felipe Cardoso landfill. The cooperative consisted of two dozen homes constructed by the residents themselves, many of whom had formerly lived in the Villa del Cerdito (Pig Town) shantytown located on top of the old Usina 5 landfill. COVIFU was divided into two strips of homes separated only by a row of sports pitches, a patch of wood, and an after-school club attended by children from the cooperative and wider neighbourhood. On one side was COVIFU Rural, officially situated in rural Montevideo, and on the other was COVIFU Urbano, in the urban part of the city. In 2010, I had been recruited as a labourer on the COVIFU building site, getting to know the future residents and never guessing that I might one day come to occupy one of the homes myself.

All the houses officially belonged to the cooperative rather than to individual residents, but the COVIFU Rural house that I was given to live in had been assigned to social worker and self-proclaimed 'lay missionary' Óscar, who had yet to occupy it. In exchange for staying rent-free for the year of 2014, my partner Mary and I agreed to fix up the house, a shell that initially lacked a toilet, furniture, or even windows. Such 'house-work' was also an opportunity to ask a series of questions of my new neighbours, including the type of goods that could be recovered from the landfill and deployed in the home and the kind of life that could be lived in and around the landfill. As I discovered, rather a lot could be extracted from the landfill or intercepted before reaching it, including my bed, mattress, cooker, and even bathroom door. As for my neighbours, almost without exception, they let me into their lives, inviting my partner and I to family parties, 'tropical music' concerts, children's football tournaments, church services, and Afro-Uruguayan religious *sesiones*. I was even taken to cock-fights, although my own cock, Wallace, unfortunately never made it into the arena, despite being fed on a diet of porridge oats and massaged with whisky.

My fieldwork site brought me into contact with waste-pickers who had diverse workplace locations, collective labour arrangements, livelihood strategies, and degrees of economic (in)formality. I did not so much follow the waste or trace the 'social life of things' (Appadurai 1986) as follow my neighbours, joining them waste-picking at a range of sites. Initially, I spent most time with neighbours Stevie, Michael and Victoria at COFECA, but on occasion I also joined another neighbour, Morocho, who worked with the Pedro Trastos cooperative. I accompanied my immediate neighbour, Morocho's stepson Juan, when he recovered materials at the landfill. Towards the end of my fieldwork period, I asked Natalia if I could work at the recycling yard that she ran with her adult children, most of whom were neighbours in COVIFU. Finally, when COFECA was disbanded and its workers were incorporated into a new formal sector recycling plant, I followed them to the conveyor belt of the Planta Aries. As well as working with neighbours, I conversed informally over innumerable *mates* and interviewed them about their lives and labour, hopes and dreams.

In order to understand the materiality of rubbish and waste-pickers' orientation towards it, I became an apprentice in waste-work, building on my amateur activity started back in Scotland. A willingness to get my hands dirty was often perceived by waste-pickers as proof of my humility, a trait which might well be considered Uruguayans' 'paramount value' (Dumont 1986). The importance placed on humility is one reason why the Uruguayan president at the time of my fieldwork, Pepe Mujica, touted by the BBC as the 'world's poorest president' (Hernandez 2012), was celebrated even by political opponents for donating 90 per cent of his presidential salary and continuing to live in a humble country home with his wife, three-legged dog, and Volkswagen Beetle. When I asked 'The Ant', a veteran waste-picker with a formerly fearsome reputation, how I would have fared back when the dump was a tougher place and knife-fights were common, he said that I would have been treated well because I was humble, 'and if you act with humility, you are treated with humility'.

Physical and sensorial engagement in waste-labour also helped me to avoid the reification of waste as a uniform category (Gille 2010: 1050), an 'amorphous blob' (Van Loon 2002: 106), or merely the discards of cultural classification. As Reno (2015: 159) notes, 'nothing is waste in general but only in particular'. The waste-picking trade involves a range of skills and sensorial sensitivities, while the move into recycling plants requires training in what Carenzo (2016) has called a 'craft-in-the-making'. Adoption of the anthropologist-apprentice model (Downey et al. 2014), meanwhile, was inspired by Ingold's (2000) argument that it is by engaging, moving, and working together in a shared environment that we come to understand and perceive the world. Participant observation, in this framework, 'allows the ethnographer to access other people's ways of perceiving by joining with them in the same currents of practical activity, and by learning to attend to things – as would any novice practitioner – in terms of what they afford in the contexts of what has to be done' (Ingold 2011: 314). Bowker and Leigh Star (1999) argue that classifications and categories are often concealed. In this case, both classifiers and their classifications are often hidden from the general population and ethnography can therefore also be grasped as a 'tool for seeing the invisible' (1999: 5).

In the UK, the United States and much of Europe, citizens engage in daily acts of domestic classification as they separate food waste from packaging, dry from wet, recyclables from plain old rubbish. This kind of separation is also hidden away in domestic spaces, with containers stashed under sinks and material stockpiles taken 'out the back'. Yet this domestic classification is only the beginning of the story, the initial first step that enables subsequent, finer processes of classification as metals are subclassified into copper, steel, and iron; plastics become PET, high-density polyethylene (HDPE), and polyvinyl chloride (PVC); and paper is subdivided into various categories according to type and quality. When we purchase, as is increasingly common worldwide, products that proudly boast that they are made from a certain percentage of recycled plastic, wool, or paper, that choice is ultimately only possible due to prior, obscured and often unrecognised processes and politics of classification, generally carried out by waste-pickers in the Global South, and by households and high-tech processing plants (also known as materials recovery facilities or MRFs) in the Global North.

One of my first jobs as an apprentice in Uruguay was to explore the classificatory categories employed by my waste-picking interlocutors. These broadly divide up the waste-stream into *material*, *requeche* and *basura*. *Material* corresponds to the different categories of materials that they recover in order to sell to intermediaries by the kilo. This is then subclassified into *blanco* (white paper), *botellita* (PET bottles), *pomo* (high-density polyethylene), *metal* (non-ferrous metals), *hierro* (scrap metal), *cartón* (cardboard) and *nylon* (low-density polyethylene). The labour of selecting *material* can be understood principally as 'commodity classification' (O'Hare 2013): clasificadores generally know the market price of such materials and their recovery involves sensory skill in identifying valuable items. The items then make their way along recycling chains that often exceed national borders. One Uruguayan buyer sent plastics contraband to Brazil's southern states, while another told me that he exported to China. Other destinations were national. Paper went to the IPUSA plant that converted scrap paper into new products like nappies, tissues, and sanitary products, while non-ferrous metals eventually made it to Werba, and scrap metal to a steel forge on the outskirts of Montevideo owned by Brazilian industrial powerhouse Gerdau.

Classifying *material* can be contrasted with the work involved in the recovery of *requeche*. While a large part of clasificadores' time was spent sorting plastics, paper and metals, they would also invariably make a little pile of interesting things to be taken home or sold at market. This generally included an array of heterogeneous things, including packets of spices, children's toys, empty containers, electronics, ornaments, and clothes. There was, in the process of selecting such items, rather more creativity, spontaneity and, I would argue, non-human agency involved, a 'thing power' (Bennett 2010) whereby materials exert an attraction over and above human intentionality. Finally, we have the category of *basura* (rubbish), used by clasificadores to describe the waste-stream generally but also contrasted with the previous terms, so that a dump truck without much *material* or *requeche* might be dismissed by clasificadores as 'pure rubbish' (*pura basura*).

An ethnographic apprenticeship in waste-work was more than just a methodological matter. Like Ingold, I reject an understanding of social life where different cultural representations are imposed on the world 'out there' in favour of the idea of collaborative, fluid, but also political classifications made by subjects-in-their-environment. Mary Douglas's (2002 [1966]) claim that dirt was 'simply matter out of place' has become widely accepted as underlining the culturally relative, as opposed to universal, nature of ideas of purity and pollution (for alternative frameworks, see Liboiron 2019; Reno 2014). Yet as Strasser (1999: 9) has argued, different conceptualisations of waste often have as much to do with class as with culture. This book suggests that class, understood as both a social construct and the relationship of subjects to the productive economy and its leftovers, is key to explaining different approaches to waste. Rather than belonging to a culture that classifies the environment in ways that an ethnographer (from a different culture) could never access, clasificadores classify waste by drawing on sensory immersion, imagination, and attunement to the political economy of discards, similar affects and modes of knowing to those exercised by dumpster divers in the Global North. By tearing open rubbish sacks, picking over waste piles, gathering expectantly in front of overflowing dump trucks, and selling materials to intermediaries, I came to understand how the heterogeneous materials officially bundled together under the categories of household, industrial, and commercial waste were perceived by waste-pickers. I also began to see these materials in a new conceptual light.

Common wastes

Paraphrasing Bruce Robbins' (2007) assertion that infrastructure 'smells of the public', we might say that urban waste smells of public infrastructure. The life of waste begins when mixed materials are placed inside municipal waste containers that constitute a 'crucial interface between waste infrastructure and waste practices' (Metcalfe et al. 2012: 137). An encounter between the citizen and the state, the act of putting out the bins is also a moment when private discard becomes public waste. As Italo Calvino (2009) puts it, the bin 'proclaims the role that the public sphere, civic duty and the constitution of the polis play in all our lives' (2009: 67). In many places in the world, materials then become 'public' in the sense that they are managed by a public entity – the local council or a private concessionary financed by the public purse. To discard waste is to 'situate it in the channels and protocols of waste management' (O'Brien 2007: 203). Such materials become public in another sense too, as disturbed bins and unruly rubbish become a matter of public debate and concern (see Latour 2004). We might recall the comedian Bill Hicks' (1997) skit in which he teased his audience about arriving in England after the LA riots and finding the newspapers ablaze with the story of hooligans having 'knocked over a dustbin in Shaftesbury'. But the property status of waste is no laughing matter for those who depend on it for a living, and the kinds of claims that Uruguayan waste-pickers make on waste suggest a particular type of *res publica*: a commons.

The idea that waste might be considered as a commons first occurred to me when I pondered my friends' relationship to the landfill. 'You can get anything you need from the *cantera*' – Matute told me soon after I had started working with him at COFECA – 'clothes, food, furniture, building materials, whatever'. He then took pride in kitting me out with a t-shirt, boots, and trousers that he fished out in just one afternoon's labour. The word used by clasificadores to designate the landfill – *cantera* or quarry – suggests a relationship of extraction, while their nickname for it – *madre* or mother – indicates one of care and provision. When I interviewed an older informant, Selva, she told me of a time when the *cantera* was *libre*, when the landfill's bounty was 'free' in the sense of being recoverable without cost and waste-picking labour was also free of harassment or interference from municipal workers or police. This book will argue that the *cantera libre* was, like the historic rural commons, a vital space to which vulnerable subjects could turn for a livelihood, food, and shelter in times of need. When Juan, my best friend and neighbour in the waste-picker housing cooperative, led me through a hole in the fence of Montevideo's landfill, and older clasificadores told me of historic struggles against the securitisation of waste, I was immediately reminded of the rallying-call of England's commoners: 'Down with the fences!'

Traditionally, the wastes that have been discussed in relation to commons have tended to be either wastelands used as commons, or commons enclosed on the basis that they are wasteful (Gidwani and Reddy 2011; Goldstein 2013; Locke 2005 [1689]). Only a few scholars have tentatively toyed with the idea that modern discards, the mass waste of consumer society and urban living, might be considered an urban commons (e.g. Negrao 2014; Zapata and Zapata Campos 2015), while others have focused on processes of enclosure and dispossession vis-à-vis waste (e.g. Inverardi-Ferri 2018), or customary rights to access refuse (Butt 2019). This book argues that 'rubbish' comes into being when materials constitute a domain of zero-value for those who dispose of them, simultaneously reaffirming the 'positive valuation of bodies and spaces as clean' (Hawkins 2003: 41; Whitson 2011). Economic models often account for waste in terms of absence – either the result of zero inputs or entailing zero cost – something which has been critiqued by those pointing out that the production of value entails the simultaneous, and hardly cost-free, production of waste (Gille 2010). Yet in an important sense waste does holds zero-value for its disposer, even as it retains potential that might still be exploited by others. As this book will make clear, the potential of discarded materials cannot be restricted to transformation into commodities (Appadurai 1986; Whitson 2011) but must also account for their potential to constitute relations of care, intimacy, and patronage.

Curiously, however, when waste is recategorised as resource, this is, for waste-pickers, the precise moment when a material ceases to be a commons. Although clasificadores do effectively treat waste-stream materials as resources, access to them depends on the fact that the previous owner has, in classifying them as waste, abandoned them. When businesses and individuals begin to treat waste materials as resources, I suggest, they often start to commercialise them as commodities, depriving waste-pickers of their livelihoods. My position thus differs from that of Bruce Lankford (2013), for whom waste only becomes a fully fledged commons when it ceases to be waste and is converted into a resource. As I will explore in more detail later, zero-waste and circular economy schemes can, alongside modern waste management technologies and environmental services, paradoxically serve to dispossess waste-pickers in the Global South and hinder other forms of discard recovery in the Global North.

This book is characterised by an 'ethno-historical' approach that combines ethnographic and historical sensitivities to the characteristics of particular commons. In McCay and Acheson's (1987) anthropological volume on the commons, Pauline Peters argues that 'opposition to the tragedy of the commons has bred its polar opposite: romanticised notions of a precommercial, precapitalist past when communal rights preserved the land and permitted all to use it on an equal footing' (1987: 177). Similarly, David Harvey (2011: 101) critiques an approach 'typically laced with hefty doses of nostalgia for a once-upon-a-time, supposedly moral economy of common action'. But there is nothing fairytale-like about the detailed social history carried out by historians of the English commons like E.P. Thompson (1991), J.M. Neeson (1993), and Peter Linebaugh (2008: 2014). Montevideo's waste commons has, I suggest, much in common with the 'classic' case of the English commons and I thus draw on participant observation and oral histories conducted with clasificadores to make a historical comparison. Articles, volumes, and even whole sub-disciplines have emerged in response to Garrett Hardin's (1968) article on the 'tragedy of the commons' but few have noted that Hardin borrows his argument from a Malthusian propagandist for English enclosure (Thompson 1991: 107). I opt to return to the roots which sustain later commons scholarship, foregrounding the heterogeneous English commons in order to illuminate the contemporary predicament not only of clasificadores, but of millions make a living from discards worldwide.

As Thompson (1991: 151) cautions, English 'common right is a subtle and sometimes complex vocabulary of usages, of claims to property, of hierarchy and of preferential access to resources, of the adjustment of needs, which … must be pursued in each locality and can never be taken as "typical"'. Nevertheless, from the work of social historians, we can draw out some shared characteristics of the English common territories. They were landscapes from which commoners extracted use and exchange value; they were an invaluable resource for the poor; and they were a domain particularly associated with vulnerable subjects. Enclosures forced commoners into migration, the poorhouse, or the ranks of the proletariat (Hammond and Hammond 1987 [1911]) but were resisted during centuries by those who asserted their rights in the courtroom and through the destruction of enclosure's infrastructure ('Down with the fences!'). Drawing on Thompson, my comparison is largely with the English commons of the eighteenth and nineteenth century but Scottish and Irish territories underwent similar processes. As Linebaugh (2019) argues, occupied Ireland acted as a laboratory for land reform policies of the British state in the eighteenth century, leading to enclosures, the replacement of peasants with livestock, and bitter resistance movements such as that of the intriguing Whiteboys Rebellion. Tom Devine (2018), Scotland's pre-eminent historian, meanwhile, has recently written of the 'forgotten history of dispossession' in the Scottish borders, before the better-known events of the Highland clearances. There, a reliable contemporary observer, Sir John Clerk of Pennicuik, wrote that the population of Galloway had been 'much lessened since the custom of inclosing their grounds took place' (in Devine 2018: 100). I will argue that there are deep resonances between processes of enclosure and dispossession that occurred in the eighteenth century and are occurring in the twenty-first.

As in the slogan espoused by the clasificador children in El Cerro, customary claims over the English commons were not generally based on the ideas that commons should be open access or available to everyone, but on the idea that they 'belong' to a particular group of subjects. As we shall see, the poor, and more specific vulnerable groups such as single mothers, recent immigrants, ethnic minorities, and neighbours living close to the landfill, are those who assert rights over the waste-stream in ways analogous to the rights claimed by other vulnerable groups to traditional commons landscapes such as fields, forests, and marshes. Another continuity between the waste and rural commons has already been mentioned: Montevidean clasificadores obtain clothes, food, shelter, and fuel from the waste-stream in ways similar to how the English landless poor used the traditional commons for subsistence. To take the case of just one type of commons, that of the open field, an English Midlands observer wrote in 1767 that:

> little parcels of land with a right of commons of a cow or three or four sheep, furnished them with wheat and barley for bread ... with beans or peas to feed a hog or two for meat; with the straw they thatch their cottage, and winter their cow, which gives breakfast and a supper of milk. (Thompson 1991: 176–7)

Now consider my neighbour Juan, who brought home from the landfill food for his pigs and sheet metal to roof his stable. His children's breakfast of yogurt was provided not by his own cow, but from the recovered leftovers of the national dairy cooperative, CONAPROLE. Like the rural commons, reliance on and access to the waste-stream also provides a refuge from wage labour, a point explored in greater depth in chapter 2.

The final parallel I wish to point to is not a feature of the commons per se but rather a contrary process: enclosure. Amin and Howell (2016: 14) have warned against 'misleadingly straightforward dichotomies' such as that between commons and enclosure, and I have no interest in resurrecting one here. Instead, I set out to chart varied processes of commoning and enclosure, where the term 'enclosure' covers the many ways that the waste commons are denied to the poor. These include what I call 'hygienic enclosure'. By this I mean, first, the material and disciplinary technologies used to construct a sanitary landfill and street-level containment of waste. For example, this might include the closure of a dump where waste-pickers could previously access materials freely and its replacement by a sanitary landfill operated by a multinational company and staffed by private security. But I also use this term to mean the legislative attempts to re-channel materials and the interception of goods when they occupy a liminal state as surplus to particular requirements, but yet to be converted to waste. Here we may think of the surplus food stocks locked in supermarket bins, unable to be accessed by or distributed to vulnerable groups, a common state of affairs that, as we shall see later, has begun to be rolled back only in a limited fashion by so-called 'Good Samaritan' laws.

Thinking about modern, mass waste as a commons also allows for a renewed exploration of the link between commons and territories designated as 'wastes' for their supposed lack of exploitation or inappropriate use. Foundational liberal thinker John Locke (1993 [c.1681]), argued that the underuse or misuse of land was justification enough for its appropriation. For Locke, 'land that is left wholly to nature, that hath no improvement of pasturage, tillage or planting, is called, as indeed it is, waste' (1993 [c.1681]: 282). But land where fruit is left to rot on the branch or vine, 'notwithstanding ... enclosure, was still to be looked upon as waste, and might be the possession of any other' (1993 [c.1681]: 276–7). Thus, as John Scanlan (2005: 24) concludes, 'in Locke's terms, claims to property ownership rest on an idea of the proper use of land, which entails the appropriation (through the use of one's labour) of its previous unused potential'. As has been shown, Locke's theories were not confined to the ivory tower but were bound up with the practices and institutions of colonialism: he played a key role, for instance, in securing British ownership of the plantation settlement of Carolina, which displaced the native Cherokee and other indigenous peoples (Arneil 1994). Arneil argues that Locke's support for colonial expansion helped to transform what was in fact a minority view, with his contemporaries concerned that colonialism would be a drain on the nation's finances, as well as diminishing its population.

Dispossessions of commons justified by under-productivity or over-use arguments have often provided fuel for discussion and rebuttal (see Chibnik 2011). Anthropologists have engaged with what have been called the 'big five' topics of commons research (forestry, fishing, animal husbandry, water management, irrigation), and have joined others in attempting to demonstrate the ecological effectiveness of commons management. The issue of conservation is rather complex for the urban waste commons, since waste appears to be ever-replenished and thus less in need of regulation (see Zapata and Zapata Campos 2015: 98). A landfill manager tells Josh Reno (2016: 4) that 'the garbage keeps coming', while Brazilian *catadores* (waste-pickers) tell Kathleen Millar (2014: 39) that it 'never ends'. The daily arrival at the landfill of fresh waste means that a tragedy like the depletion of arable pasture, fish stocks or forests cannot easily occur. This is not to say that waste or its various material fractions is an infinite resource – only a certain amount of copper arrives at Montevideo's landfill every day – more that dispossession rather than depletion was a greater concern.

Instead of conservation, this book is thus more firmly situated alongside the work of what Craig Johnson (2004: 408–9) calls 'entitlement scholars': those who 'emphasize the historical struggles that determine resource access and entitlement, and the ways in which formal and informal rules create and reinforce unequal access to the commons'. Clasificadores' claims to Montevideo's discards are made from a position of supposed weakness: their location outside of wage labour. In most Latin American and developing countries, informal sector waste-pickers have long been the primary actors in the recycling trade, with formal sector firms making an appearance higher up the supply chain and the state only in recent years. In effect, the evolving relationship between the spontaneous 'commoning' of waste by waste-pickers and its circulation within a formal state recycling scheme moves in an opposite direction to developments in European infrastructural provision, where the state has in recent years receded, allowing commoning initiatives, as well as the private sector, to plug infrastructural gaps (Dalakoglou 2016). In Latin America, the commons are often not emergent, but to be defended. The waste commons in particular, I argue, is threatened by the entry of multinational capital and its state partners into the recycling landscape. This book seeks to bridge the gap between critiques of such multinationals, the social science of waste, and interdisciplinary commons research through an ethnographic approach that highlights the status of waste as an urban commons, the role of Montevidean waste-pickers in infrastructural provision, and the dispossession-by-differentiation they face through the contemporary modernisation of waste management.

Beyond Uruguay

According to the International Labour Organization (ILO 2013), between 19 million and 24 million people globally rely on waste to make a living. On 1 March 2020, to commemorate Global Waste Picker Day, researchers working with pickers released a thematic map of socio-environmental conflicts in the Global South related to informal recyclers. Introducing the resource, Demaria and Todt (2020) note that 'waste, once a commons of the poor, is rapidly being converted into a commodity' as public authorities 'implement new waste management models that are capital intensive and technology-driven at the cost of more socio-ecologically sustainable alternatives provided by waste-pickers'. In total, the map includes over 50 conflicts in Africa, Asia, and Latin America and divides threats to waste-picker livelihoods into those that involve incineration, privatisation, and urban space restrictions. Of the mapped cases at least a third involve the closing, privatisation, or repression of waste-pickers at landfills. This has occurred to places as far apart as Panama, Brasilia (Brazil), Dakar (Senegal), Lagos (Nigeria), Nairobi (Kenya), Johannesburg (South Africa), Phnom Penh (Cambodia), Delhi (India), Addis Ababa (Ethiopia), and Qalyubia (Egypt).

'If someone plays with my [daily] bread, then I know what I will do.... I have nothing to lose but my [waste] carrier.' These words, spoken not by a Montevideo clasificador but by a Turkish waste-picker, allude to a common exposure to threats over customary waste access (Dinler 2016: 1846) and inevitably remind one of Marx's famous rallying-cry of workers having 'nothing to lose but their chains'. Waste-pickers worldwide, it seems, are at risk of losing access to waste and the tools that they use to collect it, from horses and carts confiscated in Montevideo to 'carriers' appropriated by the police in Istanbul. Indeed, attempts have been made to unite waste-pickers regionally and globally through organisations such as the Network of Latin American Recyclers (Red-LACRE) and the Global Alliance of Waste Pickers. Assertions that 'the revolution will be a long time coming to the garbage dump' (Birkbeck 1978: 1881) and that 'in most cases, waste-picking and recycling does not have a proper labour organisation' (Choudary 2003: 5242), no longer hold true.

Nevertheless, the class status of waste-pickers and their role within recycling economies has been disputed. Intriguingly, ragpickers appear centre stage both in pre-Marx bourgeois categorisations of the proletariat and in Marx's descriptions of the lumpen (rag) proletariat. In dictionary definitions that pre-date Marx, the proletariat were defined as those who produced offspring, not value, 'mean, wretched and vulgar', a heterogeneous mob or the 'dangerous classes' including 'the ragpicker and the nomad' (Frégier 1840; Thoburn 2003). Then, in one of the most famous passages of *The Eighteenth Brumaire*, Marx (1975 [1852]) firmly positioned ragpickers within Paris's lumpenproletariat, part of a motley crew that also included tricksters, escaped galley slaves, vagabonds, and brothel keepers. Indeed, Nicholas Thoburn (2003: 5) argues that the latter term was 'Marx's mechanism for freeing up his concept of the proletariat from the bourgeois image of a seething rabble', a move which 'transfers all the old content into the new category of the lumpenproletariat'. In order for the working class to be instituted as the only pure and purifying agent of emancipation, this institution had to produce its own excremental excess.

To some degree, such a displacement also served to racialise the working class. It should be remembered that the Irish, the lowest paid workers in Victorian Britain and often classified as lumpenproletariat, were not regarded by employers as sharing the same race as their British counterparts but as forming part of an inferior 'Irish race', undeserving of the same wages as the British (Robinson 2000 [1983]). Indeed, such a state of affairs continued well into the twentieth century, where certain Glasgow shipyard jobs would have been off limits for my Irish-Catholic grandfather. What Cedric Robinson calls 'racial capitalism' emerged from, rather than broke with, the racialisation that took root in feudal structures, the 'rationalization for the domination, exploitation, and/or extermination of non-"Europeans" (including Slavs and Jews)' (2000 [1983]: 27) that preceded modern nationalism. As Robinson argues, nor were the effects of racialism confined to the bourgeoisie: they found their way into the ideas of radical intellectuals and the works of Marx himself, creating an understanding of the universal revolutionary subject that went on to be critiqued by those writing in the black radical tradition. In Uruguay, these debates are relevant not only because Afro-Uruguayans are strongly represented within the waste-picker population but also because, unlike in other Latin American countries, whites are too. Across Latin America, the transformation of waste-pickers into an organised labour force has undermined their racialised characterisation as a lumpenproletariat but white waste-pickers in Uruguay are also racialised as 'los negros de la cantera': the blacks from the dump. Intriguingly, some white waste-pickers adopted a variant of this narrative, telling me that the dirt from the landfill 'turns all of us black', a point which echoes Kathleen Millar's (2020) analysis of racialisation at the Gramacho dump and the representation of waste-pickers in the 2010 film *Waste Land*. For Millar, not only waste-pickers but also the jobless poor and destitute more generally are often perceived as 'not quite white or are associated with blackness regardless of individual phenotypical attributes' (2020: 9).

Stallabrass (1990: 84) emphasises that for Marx and Engels the lumpenproletariat were not *part* of the proletariat, not least because they lacked a mutually constitutive relation with the bourgeoisie. Surprisingly, however, there is little acknowledgement from Marxian theorists that pickers, then and now, form a crucial first link in a chain of economic production.

More recent scholars do attempt to insert waste-pickers into the working class. Birkbeck, for example, argued in a pioneering article in 1978 that the Colombian landfill waste-pickers with whom he conducted research constitute 'self-employed proletarians in an informal factory'. Although they considered themselves autonomous operators, Birkbeck argues that they were in effect 'little more than casual industry outworkers' operating under the 'illusion of being self-employed' while simultaneously selling their labour power (1978: 1174). As other waste-picker scholars have suggested, at first glance this seems counter-intuitive: surely waste-pickers sell recovered waste as a commodity rather than their labour power? Yet, as Dinler describes for Ankara, waste-pickers regularly engage in various forms of social and material labour in order to access waste and transform it into a viable commodity.

Broadly then, two fundamental categorisations of waste-pickers' class position can be identified: proletariat and lumpenproletariat. Yet in practice these are not so far apart as they might seem. Echoes of the idea that waste-pickers form part of the lumpenproletariat could still be found in Uruguayan clasificador politics, from Communist Party technocrats in the Ministry of Social Development (MIDES) to anarchist activists in the UCRUS. The latter was affiliated with the national trade union federation (the PIT-CNT), implicitly recognising waste-pickers' status as workers. Yet the union was also not averse to supporting more explicit proletarianisation, including the construction of a recycling mega-plant that would have provided employment for 500 clasificadores. As Stallabrass (1990:70) notes, the connotations of the word lumpen – rags, tatters, shabby – 'suggest less the political emergence of a class than a sartorial category'. All the more fitting then that MIDES, as we shall see, sought to clothe clasificadores in the uniform garb of the proletariat, as well as structure their habits and consciousness with fixed working hours, salaried employment, and a management structure. Yet to some degree, such plans only sought to clarify an existing economic relationship between industry and waste-picker, cutting out the middle-man and bringing Birkbeck's 'industry outworkers' in-house.

In this book, I follow Kasmir and Carbonella (2008: 6) in challenging the division made between the proletariat and the lumpenproletariat and the working class and the poor, which often maps on to a geographical division between Global North and South respectively. Rather, we take as our starting point the commonalities to be found between salvaging practices in the Global North and South to explore the discursive and material processes of proletarianisation in which waste-pickers are implicated. These include demands for recognition, enrolment in cooperative schemes, recruitment as public and private sector workers, organisation into trade unions and mass movements, and defence of customary access and ways of life, including kin- and ethnicity-based organisation of labour. A basic conflict underlies many of these schemes. Can waste-pickers, and indeed dumpster divers, be considered as, and do they consider themselves to be dignified workers, even when dressed in rags, circulating on horseback, sticking their heads into bins, and knee-deep in mud at landfills? Or do they need to be dignified through collectivisation, proletarianisation, and inclusion into modern waste management systems?

As intriguing as these questions are, they can often prove a moot point, because what is on offer for most of the world's waste-pickers is neither the status quo nor a place in a formal waste management system. For many, customary access to waste is under threat from multinational companies that offer their services as sanitary landfill managers, provide recycling services or, increasingly, seek access to waste as a feedstock for waste-to-energy plants. Crucially, and as I will explore in chapter 4, this dispossession often involves a fracturing of waste-pickers as a group that is reminiscent of the proletariat/lumpenproletariat division. This new double movement excludes as it includes, offering working-class jobs for the few in modern recycling supply chains or waste management systems while disenfranchising the many who subsist, and occasionally thrive, through the commoning of capitalist excess. Rather than the spectre of accumulating wastes so familiar to the environmental imaginary, it is rather their disappearance – as they are transmogrified into resources, energy, and feedstock – that most concerns us here. At risk is not just the political economy of recycling but also its moral economy, within which waste-pickers – from Uruguay to Istanbul – are at pains to distinguish themselves from their erstwhile lumpen bedfellows: tricksters and drug-dealers, prostitutes and thieves.

In practice, Brazil and Argentina are the countries with which flows of Uruguayan waste, ideas, and activism are most directly connected. In these three countries, a series of interlinked economic crises took place between 2001 and 2002. Waste-pickers, known in Brazil as *catadores* and in Argentina as *cirujas* or *cartoneros*, certainly pre-date these crises (Suárez 2016) but the sharp economic downturn made thousands of workers unemployed, and many of these turned to recovering materials from the waste-stream in order to earn a living. In 2003 in Uruguay, for instance, an 'obligatory' but not legally binding Intendencia census recorded the number of clasificadores as having doubled to 7,200 (Chabalgoity et al. 2004: 14). Although consumption and thus the generation of waste slumped, the devaluing of local currencies also made imports more expensive, driving up the price of local recyclables and making this form of livelihood more viable. The increased number of waste-pickers translated into their heightened visibility and, for many, such figures came to be conceived of as symbols of national crisis, a re-emergence of the lumpen in countries with strong traditions of work-based citizenship (Grimson 2008; Whitson 2011).

Increased media and middle-class interest in waste-pickers in the 2000s was accompanied by a growing scholarly output. This work has focused on issues such as the economics of the 'informal' waste trade; the stigmatisation and discrimination suffered by waste-workers; the history of waste-picking; and attempts by waste-pickers to organise into cooperatives, trade unions, and political movements. The latter organisation of waste-pickers was encouraged by a series of developments. First, a significant number of workers who were made unemployed brought their trade union experience into the waste-picking trade, and some political activists became clasificadores precisely in order to organise the sector (see chapter 5). Second, a variety of non-governmental organisations (NGOs) accompanied and encouraged the collective organisation of waste-pickers, ranging from local organisations to the continent-wide philanthropy-financed Avina (Rosaldo 2016; Sorroche 2015). Third, the election of a series of centre-left governments in Latin America – the so-called 'pink tide' (see González 2018) – created an atmosphere propitious to the inclusion of organised waste-pickers into state waste management programmes (Marello and Helwege 2014).

These developments in no way constituted a uniform process. Important differences exist between countries, cities, and indeed municipalities: the fact that waste management is almost always devolved to local authorities means that radically different approaches towards waste and the waste-picking 'problem' can exist in neighbouring localities. Nevertheless, in Argentina and Uruguay we can identify a common trend towards the decriminalisation of waste-picking (occurring in Montevideo in 1991, and in Buenos Aires in 2002) and the inclusion of waste-pickers into municipal recycling plants (in Buenos Aires from 2003, and in Montevideo in 2014). Brazilian President Luiz Ignácio Lula da Silva (2003–11) was sympathetic to *catadores* and often addressed their national congress, while

President Dilma Roussef's (2011–16) government brought in groundbreaking national legislation to favour waste-picking cooperatives in municipal waste tenders. While the broader Latin American panorama is expectedly diverse (see Red-LACRE 2017), Colombia deserves special mention for its historic waste-picker trade union, its levels of cooperativisation, and the role that waste has played in recent public life (Rosaldo 2016; Samson 2015a; see also Birkbeck 1978).

The vast majority of the world's waste-pickers do not work in cooperatives or recycling plants and are not active trade union members (Medina 2005). Most academic research, however, has been conducted with such collectives, shaping the profile of waste-picker scholarship. There is much of value in this diverse body of work, and this book seeks to build on some of the central themes raised. Yet long-term ethnography conducted with waste-pickers who do not form part of an association or cooperative is sparse (although see Gorban 2004; Millar 2018; Schamber 2008), an oversight that can implicitly reproduce the Marxian assumption that most waste-pickers form part of a lumpen and non-revolutionary proletariat and are therefore unworthy of attention. I seek to redress this imbalance, showing that even where they are not employed by the state or organised in collectives, waste-pickers provide an important infrastructural and industrial service when they access the waste commons and collect and recycle materials disposed of by the population.

The labour of clasificadores, I will argue, effectively constitutes a 'shadow infrastructure' that absolves institutional waste service providers from some of their responsibilities while potentially creating new challenges for them through environmental contamination. Recent anthropological literature on infrastructure and waste has concentrated on the materiality of waste infrastructure (Chalfin 2014; Fredericks 2014; Harvey 2013; Miraftab 2004); its temporality; the relation between the flow, interruption and visibility of infrastructure (Dalakoglou and Kallianos 2014); and the potential of waste infrastructure to stimulate new politics and publics (Chalfin 2014; Stamatopoulou-Robbins 2014). In focusing on the Intendencia's discursive and material treatment of Montevideo's waste, chapter 2 speaks most directly to the anthropology of infrastructure (see also Anand et al. 2018; Larkin 2013). However, the wider argument and its detailing of waste-picker practices can also be seen as contributing to these discussions if we recognise, as waste-picker activists have long argued, that waste-pickers provide an infrastructural service when they collect and recycle materials disposed of by the population. As in Fredericks' research in Dakar, infrastructure in such instances is 'devolved' onto labour and bodies, which 'bear the brunt of this labor-intensive infrastructure through the onerous physical demands of the work itself, associated diseases ... and the stigma of laboring in filth' (2014: 539). Beyond waste, this understanding of infrastructure coincides with Simone's (2004: 408) argument that people can be understood as infrastructure in many cities of the Global South and where urban African ruins mask a hive of social infrastructure. Until recently, an official infrastructure of recycling has not been present in Montevideo, only the 'spontaneous' (Fernández 2012) recycling carried out by clasificadores. This book highlights not only how waste provides a means for families to take care of each other, but also that kinship constitutes the backbone of a social infrastructure of recycling in Uruguay and beyond.

Beyond the suffering scavenger

My engagement with waste in the UK and Uruguay invites a rethinking of waste-picking labour, generally conceived of as abject, stigmatising, and hazardous. As Josh Reno has highlighted, the news media, for example, 'often uses scavenging as an index of global inequality' (2009: 32). Joel Robbins (2013), meanwhile, has argued that anthropologists have replaced the 'savage slot' with the 'suffering subject', so that 'the subject living in pain, in poverty, or under conditions of violence or oppression now very often stands at the centre of anthropological work' (2013: 448). With article titles such as 'amidst garbage and poison', the work of Auyero and Swistun (2007) on 'polluted peoples' is representative of a genre that seeks to draw attention to the environmental suffering endured by marginalised groups. Depictions of waste-pickers as suffering victims are not necessarily always negative: such representations can open up spaces for redemption and be leveraged by waste-pickers to fight for improvements in their lives and working conditions (see Alexander and Reno 2012). Like anyone's lives, however, those of clasificadores were complex and nuanced and any description of them as simply miserable (or conversely, as overwhelmingly joyous) would be incomplete and unethical.

As far back as the work of Henry Mayhew (1968 [1851]) on *London Labour and the London Poor*, it has been clear that there exists much diversity within scavenger professions, even within a single city. Over several years, Mayhew got close to the many trades exercised by the poor and working class in Victorian England, often lobbying through his newspaper articles for an improvement in their conditions. Among the waste-related professions that he categorises can be found bone-grubbers, mud-larks, pure-finders, dredgermen, and sewer hunters, with such titles bearing witness to lost vernaculars, trades, and economies. Certainly, many of Mayhew's descriptions feature 'suffering scavengers' and the squalor, inequality, and pestilence of Victorian London. Yet, in an echo of the rhetoric of Uruguayan waste-pickers, many preferred 'liberty and a crust ... to the restrictions of the workhouse' (1968 [1851]: 138). Also, like Uruguayan waste-pickers, the so-called 'sewer hunters' that rescued objects (principally metals) from the sewers that flowed into the Thames were generally proud of their good health and resistance to illness, with Mayhew noting that they were generally 'strong, robust and healthy men' many of whom 'know illness only by name' (1968 [1851]: 152).

Glasgow too, had its own band of sewer hunters, as an 1848 report in city's *Herald* newspaper described:

> Most people know there has long existed in London a peculiar class of persons called 'Mud-larks', who earn their living in the common sewers and drains of the metropolis, by picking up whatever chance or crime may throw into these receptacles. We were not aware, however, till the present week, that a similar class of subterranean operatives has, for a considerable time existed in Glasgow. (quoted in Tisdale 2018)

The article goes on to describe how, following the discovery of the sewer hunters, the Police and Statute Labour Committee of the city brought in unemployed workers on the Relief Fund to clean the sewers, presumably also clearing out the scavengers in the process, at least for a time. Interestingly, both the official and the shadow infrastructural waste-workers in Scotland's streets shared the same title – scavengers – for centuries, an indication that little stigma was attached to scavenging work per se (Skelton 2012). It was, however, the professionalised municipal infrastructure that prevailed. Of the four points where Glasgow's mud-larks accessed the sewers, only one – the Molindinar Burn – is still accessible, and these days it is only the occasional curious diver who plunges in, the market stalls whose dropped coins provided much of the sewer's valuables having long since disappeared (Cooper 2008).

Although separated in time and place, a continuity between these 'finders' and Montevideo's 'classifiers' is that both relied on discards for work rather than just enduring it in their neighbourhoods. Rubbish, in a certain vein of ecological scholarship and investigative journalism, is often involuntarily foisted upon the poor, minorities and people of colour whose low-income neighbourhoods become 'sacrifice zones' (Lerner 2010; see also Melosi 1995; Mohai et al. 2009; Renfrew 2009). Robert D. Bullard (1990) has explored the relation between race, class, and environmental quality in *Dumping in Dixie*, while Vivian E. Thompson's (2009) *Garbage In, Garbage Out* focuses on inter-state inequalities. The waste–poor nexus that I explore in this book is of a different order, one where rubbish is coveted rather than rejected, and where life is threatened by waste's absence as opposed to its presence. Further, emphasising the unsanitary, exploitative and dangerous nature of informal sector waste-picking can too often be used as a justification for dispossession (see Corwin 2019). Is the informal waste trade more exploitative than standardised wage labour? Is work at recycling plants always less hazardous and precarious than semi-clandestine labour at the landfill? These are important questions that this book seeks to explore empirically, and in conversation with interlocutors, rather than assume.

Generally, my waste-picking friends and informants have experienced an improvement in their living standards in recent years, as a result of multiple factors, including the relatively steady market value of certain recyclable materials and a decade of redistributionist polices from the centre-left Frente Amplio government. Many of my waste-picker friends who still live in shantytown housing are on waiting lists for a cooperative home or are in the process of building one; school attendance rates are up; and by the time I left the field, most of my neighbours even had shiny new sets of teeth thanks to subsidised dental care. It is also the case that many of the violent or disturbing episodes recounted to me by informants took place many years before, and that adults explicitly sought to avoid reproducing the harshness and hardships of their childhoods – including exposure to criminal activity, prostitution, and alcoholism – when bringing up their own children.

There is also some evidence that the landfill waste-pickers who constitute the focus of this study enjoyed a greater income than other Uruguayan waste-pickers. Prior to the creation of recycling plants, clasificadores were employed almost exclusively in the informal sector, and were either stationed at the landfill or collected waste throughout the city, using a horse and cart, a hand-cart, a bicycle, or simply a bag. These devices for collection and classification have themselves served to divide waste-pickers into 'carters' (*carreros*) and 'baggers' (*bolseros*), in addition to landfill 'crawlers' (*gateadores*). The total number of clasificadores is disputed, and sits somewhere between the 2,000–3,000 figure used by the Intendencia (see Matonte et al. 2017) and the figure of 15,000 regularly cited by the UCRUS. The number also varies according to the state of the economy, with unskilled construction workers, for instance, turning to *clasificación* in times of economic downturn. Estimates for the average clasificador monthly income have varied wildly in different studies and surveys: US$17 (Dirección Nacional de Evaluación y Monitoreo 2006), US$30 (Chabalgoity et al. 2004), US$146 (PUC 2006: 13), US$203 (Matonte et al. 2017) and US$610 (IM et al. 2012). The huge disparities between these numbers mean that they are unreliable but it is worth noting that Uruguay's 2006 Emergency Poverty Plan (PANES) recognised that clasificadores were not necessarily the poorest of the poor but required special attention due to structural issues of vulnerability and poverty reproduction. I think that it is also safe to assume that landfill waste-pickers, whom I calculated as earning an average of US$1,180 per month from the sale of *material* alone at a time when the national minimum wage was US$480, earned more than most *carreros* and certainly most *bolseros*.

Nevertheless, suffering was hardly absent from landfill waste-pickers' lives. We can think of the repressed trauma of past episodes of violence and the 'slow violence' (Nixon 2013) of environmental contamination, as well as personal tragedies grounded in socioeconomic conditions. Twenty years ago, veteran clasificador Ruso and his wife lost two daughters when the shack they were living in burnt down due to precarious, self-installed electrics. Their memories live on, in gothic script on their living daughter's tattooed arms and online public commemorations through regular Facebook posts that now also remember Ruso, who recently passed away. Only a few years ago, El Nani, one of the San Román brothers with whom I worked briefly at COFECA, lost two young daughters in an almost identical accident. Yet despite such recent tragic incidents, few denied that the material conditions of their existence had broadly improved in recent years. This did not prevent a certain nostalgia for days gone by that referenced both greater solidarity in times of adversity, and the libertinism of shantytown life, where parties could continue until the early hours without complaint from neighbours.

An advantage of living as well as working with clasificadores was that it allowed me to understand lives, choices, and experiences from their perspective. On the one hand, I encountered them outside work, getting to know them as mothers, brothers, neighbours, friends, pig-rearers, and amateur footballers (Figure 2). On the other, I could also observe clasificadores in their own milieu. Much research on waste-pickers has focused on the stigmatisation that they suffer as they collect waste in city centres and affluent neighbourhoods (Lombardi 2006; Magalhaes 2016; Neiburg and Nicaise 2010; Sternberg 2013; Whitson 2011). Such encounters between middle-class neighbours and waste-pickers are but one moment in the process of waste recovery, one which may be positive or negative, or indeed may not take place at all, as was the case for most of my neighbours, who had waste brought to them around Felipe Cardoso. Nevertheless, the appearance of waste-pickers in middle-class areas has dominated media reports and scholarship in cities like Buenos Aires and Montevideo, and has encouraged policies centred on themes of visibility, rights, and recognition.

Not everyone who lives in COVIFU, or the neighbourhood of Flor de Maroñas to which it is attached, is a waste-picker. But I only rarely heard of waste-pickers being stigmatised locally because of their occupation. I remember the predicament of Gonzalo, who coached Juan's son Iván at the Flor de Maroñas football team. Struggling to find work as an electrician, Gonzalo listened enviously to tales of the bounty Juan brought back from the *cantera* and wondered out loud about whether he might not give it a go himself. Casting an eye over his delicate frame, Juan doubted that he would be up to the job. Given that my neighbours spent most of their

Figure 2 The author with clasificadores during a football match in the centre of Flor de Maroñas

Source: Author photo, 15 December 2014.

time in Flor, and other popular/ working-class neighbourhoods, they did not regularly endure middle-class insults or a reproachful bourgeois gaze. When activists and academics speak of 'society' treating waste-pickers like the rubbish they collect, I suspect that they often, perhaps subconsciously, engage in an exercise of class-based synecdoche, where 'high society' comes to stand for society as a whole (see Ingold 1993: 212). The working-class residents of Flor de Maroñas are just as much part of society as anyone else, and it was with them that my waste-picking neighbours interacted on a daily basis, as they picked up their children from school or football practice, went to Sunday market, or waited endlessly for the single bus that served the neighbourhood.

The landfill centre

As Kathleen Millar (2012: 168) notes, 'much of the recent literature on urban poverty suggests that today's urban poor are excluded economically, politically and socially, and constitute a residual class that is superfluous to the global capitalist economy'. But like the Rio de Janeiro landfill *catadores* with whom Millar (2018) has worked, Montevideo's clasificadores are intimately connected to the broader citizenry by way of their waste. The city's built environment is literally sustained by their labour, crystallised in the steel girders made from recovered scrap metal. As salvaged materials travel the world as part of the global recycling trade, waste-pickers in countries like Uruguay and Brazil are also integrated into the flows of the global economy. In any case, I suspect that the centre–periphery dichotomy is not a particularly useful heuristic tool for understanding the lived subjectivities of my interlocutors. While *el centro* for me meant the historic heart of Montevideo, for my interlocutors the word *centro* was used for the local after-school club (Los Trigos) or the Portones shopping centre to the east. As a desirable leisure space, the relatively accessible stretch of beaches and their promenade (*rambla*) held much greater appeal than Montevideo's old colonial heart, rich in architectural grandeur but suffused with an atmosphere of gradual decline.

In many ways, the real centre of both neighbourhood economic life, and this book, is the landfill itself, just as landfill labour is central rather than peripheral to the global recycling industry. Viewed from the air, many elements of the surrounding built environment radiate out from Felipe Cardoso and owe their existence to it. There are the waste treatment facilities established around its perimeter, such as the privately owned medical waste disposal plant, and the municipal facilities that capture methane gas and leachates. There are the yards that companies use for keeping skips, and the spaces where waste-picking families intercept trucks. There is waste-picker housing, from the shantytown of Felipe Cardoso, to the housing cooperatives of COVIFU, COVICRUZ and COVISOCIAL, built to replace informal settlements. The *cantera* is the centre of economic activity but also of (largely male) sociality, stories, and dreams of redemption and progress. Inevitably, *la cantera* is a centre whose rhythms are partly determined by other centres of political and economic power: the commodity prices of the London Metal Exchange, the virgin oil price, decisions to crack down on contraband material flows to Brazil, and political policies that foment certain industries or allow deindustrialisation. For the most part, and with notable exceptions, clasificador knowledge of these other centres was limited to comments about PET being 'up' or cardboard 'very down', as information flowed locally by word of mouth. As Millar argues, then, waste-pickers are integrated into global capitalism but clasificadores have partial and incomplete ways of indexing that connection, one of which is market abstraction of interconnected local and global prices.

This book takes the landfill as its centre, travelling out from it in time and space in order that we better understand not only the relationship between Uruguayan waste-pickers and their livelihood but also processes of commoning, waste, and enclosure in late capitalism more broadly. The chapters chart the variegated processes that liberate or enclose waste, from the period clasificadores referred to as the 'free landfill' (*cantera libre*) through the repression of the (1973–85) Uruguayan dictatorship, to the Catholic-inspired social inclusion and labour formalisation policies of the 'progressive era' of the centre-left Frente Amplio (Broad Front) national government (2005–20).

Chapter 1 draws on archival research, oral history, and participant observation conducted with private and public sector 'waste managers' to explore the rationales and logics behind the management of Montevideo's waste, linking these to the British and European 'hygienic modernities' from which the city's modernisers drew inspiration. It argues that central to the city's hygienic and infrastructural modernity has been a conceptualisation of waste as hazardous, risky matter that should be contained and eliminated. The potential value embedded in the waste-stream has been downplayed in the interests of policing public health and private profit. In this chapter, waste is seen not as inert 'stuff' to be managed but as discursively and materially constituted through classificatory and material infrastructures. Clasificadores, it is argued, embody a shadow infrastructure focused on the recovery of value in the waste-stream. This shadow infrastructure is shown to have global parallels outside of the waste sector, in shared popular practices and forms of knowledge sidelined with the expansion of state and corporate forms of modernisation that claim to operate on *terra nullius*.

Chapter 2 explores practices of enclosure and commoning at Montevideo's Felipe Cardoso landfill and draws on contemporary ethnography and oral histories gathered from clasificadores. I trace the history of the 'mother dump', which began as an attempt to fill quarries with waste and eventually replaced the wage labour of the local brick industry with a new labour regime based on the unwaged extraction of value from urban discards. Hygienic enclosure soon followed, and involved the violent state repression of clasificadores both during and after Uruguay's military dictatorship. In response, landfill waste-pickers stubbornly resisted exclusion and remain there to this day. I explore how, like the English commons, Felipe Cardoso functions as a refuge outside of wage labour where vulnerable groups can source materials of use and exchange value. Finally, I draw on historical and ethnographic research into landfills and wastelands elsewhere to demonstrate how they too can operate as urban commons at the margins of capitalism and at risk from processes of enclosure.

Chapter 3 looks more closely at the human relations sustained by the waste commons, including the practices of care intrinsic to the strong bonds of kinship, and especially siblingship, found among clasificadores. In contrast to accounts that depict informal sector waste-picking as shady and degrading, this chapter provides examples of the ways in which discarded things help to constitute desirable subjectivities and ethical behaviours, bridging a conceptual division between 'kinship-as-care' and 'care-work' through a focus on care as the provision of kin-based labour. Discussing waste-work in the context of the (dis)embedding of labour in relations of capitalist production, the chapter argues that clasificador kinship forms the backbone of the recycling industry in Uruguay. This can simultaneously be seen as an enabler of profits for formal sector companies and an important way that clasificadores care for kin in precarious economic circumstances through the distribution of waste-labour.

Chapter 4 argues that the recent implementation of a Packaging Law in Montevideo and its concomitant recycling plants constitute a dual appropriation of the waste-picker logic of value recovery and a Latin American Catholic praxis of accompanying the poor. On the one hand, the new plants represent merely the latest instance of hygienic enclosure, whereby some materials and workers are contained in plants while the majority of waste-pickers are dispossessed of their livelihoods. On the other hand, the link between the waste commons and the poor is maintained but reconfigured, as alternative definitions and indicators of the vulnerable population entitled to labour in waste are laid bare. The chapter relies on ethnographic engagement conducted as Montevideo's waste-picker cooperatives were disbanded and workers enrolled into the 'new paternalism' of NGO-managed plants. It questions the logics and practices behind the collectivisation and formalisation of a minority of waste-pickers in the developing world and the way that this can create a cleavage in the working population that delegitimises and devalues the forms inherent to so-called 'informal' labour. Beyond a celebratory championing of recycling and re-use in general then, this chapter demonstrates how certain models of recycling deemed aesthetically, hygienically, and economically desirable can crowd out forms of 'recycling from below'.

Uruguay's Union of Urban Solid Waste Classifiers (UCRUS) taps into the country's rich tradition of anarcho-syndicalism and is the only waste-picker trade union in the world that is affiliated to their national trade union congress. Chapter 5 draws on longitudinal fieldwork conducted with the UCRUS to chart the tactics and methods developed by the union to defend clasificador access to waste and struggle against forms of hygienic enclosure that include the implementation of waste containers, the closure to waste-pickers of central districts of Montevideo, the securitisation of the landfill, and legal impediments to their collection of recyclables. The chapter moves beyond the landfill and recycling plant to bring in the story of Montevideo's *carreros* – horse-and-cart waste-pickers who claim a lineage going back to Uruguay's nomadic gauchos, who faced Uruguay's original enclosure or *alambramiento*. Arguing that the contemporary enclosure of waste might be considered an instance of 'urban alambramiento', the chapter describes the challenges the union faces in organising a dispersed workforce and argues for the importance of circulation as a key demand, tactic, and method for the organisation of precarious workers. Whereas conventional trade unions are involved in disputing the distribution of the surplus value, waste-picker unions, it is argued here, effectively struggle over the distribution of the surplus materials of capitalist production.

Linking the commons with the Gramscian idea of common sense recently taken up by Crehan (2016) and drawing on Polanyi's (2001 [1944]) concept of habitation, the conclusion analyses the interconnectedness of the sites of waste classification explored in each chapter, and suggests how this book's radical framing of waste, infrastructure, commons, and enclosure can prove influential beyond the boundaries of the Uruguayan field site. Theory from below and from the Global South, it is suggested, can help to reveal the exclusionary logics of capitalism behind recent moves to recategorise waste as resource and celebratory accounts of new 'circular economies'.

1

'All because We Bought Those Damn Trucks'

Hygienic Enclosure and Infrastructural Modernity

The office where I conducted fieldwork during the Uruguayan summer of 2014 is no doubt much like other public sector workplaces in Montevideo. The public servants there complain about broken air conditioning, haggle over the taking of holidays, joke and flirt with one another, and enjoy endless sips of *mate* tea in order to deal with endless amounts of paperwork. Its principal distinguishing feature was the responsibility it held for authorising and classifying the thousands of tonnes of waste deposited daily at Felipe Cardoso, Montevideo and Uruguay's largest landfill. In travelling to the Laboratorio de Higiene from my house opposite the landfill, I undertook the reverse journey of Montevideo's waste, which before reaching Felipe Cardoso had to pass through the office, in one form or another.

I had heard of the municipal Laboratorio from waste-pickers who had formalised their activity and needed municipal approval for the collection and disposal of waste. Given the institutional title, I expected to encounter a scientific environment when I arrived to interview the director. Indeed, most of the Laboratorio's small team were trained chemists who wore white lab coats. For the most part, however, they encountered waste not under the microscope, but in paper applications to be processed, approved, or declined. Under the direction of the middle-aged director Joana, a light and humorous atmosphere prevailed. After our interview, she agreed to let me return to conduct participant observation once a week. On my first day, ever keen to make myself useful, I was asked to order the office cupboard, disposing of records that had already been digitised. It was with a meandering journey through such paperwork that my archival research into the history of Montevidean waste management began, while I simultaneously noted down the queries that passed through the office. A shipload of shark meat had mistakenly arrived at the port – could it be dumped in Felipe Cardoso? A ladies' retirement home founded in the nineteenth century was updating its TV sets – would the Intendencia collect the old televisions? A private high school was replacing its scientific equipment – could they dispose of anatomical skeletons along with test tubes and Bunsen burners? I had already encountered plenty of discards as they tumbled out of trucks in and around the landfill: now I met them classified according to industry, calculated in kilograms and tonnes, and filed away under diverse rubrics.

The Intendencia is the most important actor involved in shaping Montevideo's 'wastescape', a fact equally relevant for clasificadores as it is for waste scholars. It is the Intendencia that defines and subdivides the city's waste; that stipulates what *should* happen to materials once they are thus classified; that collects refuse from most of Montevideo's citizens and grants concessions for the collection of the rest; and that owns and operates the city landfill. Specifically, the Intendencia has a fleet of around thirty public waste collection trucks, while the Consorcio Ambiental del Plata (CAP), a subsidiary of the Spanish multinational Abengoa, which holds the concession to collect waste and recyclables in the centre of Montevideo (Municipio B), operates seven trucks and four smaller vehicles (Abengoa 2014). Together, these collect waste from over 13,000 waste containers throughout the city. In their labour, clasificadores are governed by municipal decrees; they brush up against the materiality of municipal waste infrastructure at the landfill or on the streets, and they act against normative processes of linear disposal when diverting materials from the waste-stream. Rather than mere background, municipal regulations and infrastructures have important affective and political implications for the everyday labour of waste-picking.

This chapter has three central arguments that are organised by a structure that attempts to model the linear ideal whereby materials are first defined as waste, second contained, and third eliminated. At the same time, just as mixing materials is integral to waste management, we find a muddling of sequence in this model and chapter, so that infrastructures of containment and elimination designed to act on waste can in fact create it. This then, is the first argument: the classification of everything that is discarded as waste calls forth an infrastructure that attempts to prevent access to and consumption of discards by way of their destruction. Second, I sketch out the contours of what in the introduction I described as Uruguay's infrastructural modernity, arguing that its substantive feature is a process of 'hygienic enclosure', whereby discarded materials are enclosed at various scales, through the use of various technologies – containers, trucks, fenced landfills – and legal procedures, such as the establishment of municipal property rights over waste. The rationale for such manoeuvres has overwhelmingly been framed in terms of efforts to maintain the hygienic integrity of the body politic and that of the individual clasificador.

Understood within the framework of hygienic enclosure, the activity of Montevidean clasificadores is primarily seen as illicit. Beyond this optic, however, and this is the chapter's third core insight, it can also be seen as an infrastructure in its own right, albeit one that exists in the *shadow* of the state and takes advantage of waste as a commons. As 'matter that enables the movement of other matter' (Larkin 2013: 329), waste services are not like other infrastructures, such as roads allowing transport, or pylons permitting the passage of an electrical current. Instead, the collection and disposal of waste enables our economies to continue with activities of mass production and consumption. Against the ideal of infrastructural and municipal modernity then, clasificadores, with their bags, carts, horses, and bicycles, can be seen contributing to this infrastructural provision. In the final section I expand on the idea of shadow infrastructure, and demonstrate how the concept might help us to analyse processes of dispossession and modernisation outside of Uruguay and outside the area of waste. For this I make a series of 'partial connections' (Strathern 1991) that link the threats faced by clasificadores in Uruguay to those encountered by traditional midwives, gypsy scrap dealers, and informal lenders.

Classification and the creation of municipal solid waste

What is waste in Montevideo and how has this changed over time? This seemingly simple question opens up onto a series of rather more complex ones. What commodities and materials are or have been produced and consumed in Uruguay? By which processes are materials classified as waste, and according to what criteria? Can a material's classification as waste be definitive or is it always disputed? What I offer in this section is not a complete historical account, but rather a description of the major trends in the Uruguayan economy, the framework for the emergence of municipal solid waste (MSW) in Montevideo and the key distinction between waste-as-discard and waste-as-unusable-material. I begin with the changing composition of Montevideo's waste-stream before turning to the ways that municipal decrees and infrastructures not only manage but also create waste, and the ethical implications for those who act in a moral, as well as a political, economy of discards.

Wastes, in the sense of materials surplus to human requirements (Gille 2010: 1051), have been present in Montevideo since the city's founding in the early eighteenth century. To give but one example, European travellers were horrified to find cattle so abundant in the area that most of the animal was simply left to rot after having been killed. 'They do not get [from the entire cow or bull] but the leather and the tongues, which they leave to dry in the sun', commented the eighteenth-century French traveller Fesche, aghast (in Duviols 1985: 14). Briefly joined with Argentina as part of the United Provinces of the Río de la Plata, Uruguay gained its independence between 1811 and 1825 after a struggle for the territory between Argentina and Brazil was ultimately settled by British intervention (Faraone 1986 [1974]). This cemented a connection with Britain that would continue, as the rapidly industrialising superpower accounted for the lion's share of Uruguay's imports and exports into the twentieth century (1986 [1974]: 48).

Uruguay had inauspicious beginnings. Faraone characterises the post-independence country as 'a depopulated territory, [with] primitive means of production, absolute dependence on Europe for the provision of finished items, [and] abundance of products generated by the cattle industry' (1986 [1974]: 20). From cattle, the primarily rural economy diversified into sheep, whose wool supplied the European market. From 700,000 animals in 1852, the country had 13 million by 1871. Between 1870 and 1913, leather, wool, and meat accounted for 66 per cent of the country's exports (Roman 2016: 32). This provided not only a period of export-led economic growth but also, inevitably, an accumulation of wastes associated with these trades. With a relatively sparse population that mostly lived in rural areas, and without much of its own industry, the weight of the wool and cattle trade as part of an overall waste-stream would not have been marginal. The 1875–1900 period bore witness to the expansion of capitalism. By 1897, Montevideo's population had reached 280,000 and the city boasted almost 2,500 industrial establishments, of which 70 per cent were dedicated to food production (beer, sugar, flour, pasta), clothing, furniture, and construction, with glass, matches, and textiles also present (Faraone 1986 [1974]: 44). The city's landfill boomed with the products of nascent industry and a growing population, as well as those imported 'luxury goods' funded partly by economic growth and partly by debt (1986 [1974]: 53).

Municipal solid waste is an object that can hardly precede the emergence of the municipality as a political form. The work of cleaning and hygiene can be understood as activities that consolidated the authority of the state in Montevideo as the city entered a period of post-bellum stability in the late nineteenth century (see Fredericks 2014: 536). The Public Sanitation Act of 1888 was one of the first decrees issued by the fledgling city government, the Junta Económica Administrativa, making it responsible for 'extracting rubbish' from citizens, as well as a tax that would pay for the service (Fernández y Medina 1904). The decree formed part of wider municipal attempts to control the boundaries and material flows of a growing city increasingly perceived as disorderly (Baracchini and Altezor 2010). As an everyday presence, the local state became embodied in the men who picked up the rubbish as well as those who guarded the streets after dark.

For much of the nineteenth and early twentieth century, Montevideo's municipal government[1] took responsibility for materials surplus to the households located within the boundaries of the city walls. There is scarce quantitative data on the composition of household discards at that time, but this is compensated for somewhat by the rich description of late nineteenth-century waste (collection) given by flâneur, and future mayor of Montevideo, Daniel Muñoz. In the Montevidean household kitchen of 1883, he recounts, writing under the pseudonym of Sansón Carrasco:

> there is usually a rubbish box (*cajon de basura*), similar to a hospital coffin.... Affluent houses tend to have a reinforced box, presentable, decent even, if this word can be used to describe a rubbish receptacle, but the most fashionable junk used for this purpose is dilapidated kerosene tin cans that can be seen on the pavements every morning, ready for the visit of the rubbish man and brimming with all kinds of waste: rags, papers, vegetables, bones, and every type of filth that the broom gathers up during the day, from the living room to the furthest corner of the house. In the rubbish box, one can study the intimate life of each family: what they eat, what they spend, what they waste, what they save, what they work as, and what they wear. It is the index of private life, the sum of what was done yesterday, the household accounts book. If the rubbish men were observant, they would end up getting to know all the city's inhabitants intimately, finding out about their customs, their vices and their virtues, just by paying a little atten-

tion to what comes out of each box as he empties them into his carts. (Carrasco 2006 [1883]: 36)

Carrasco goes on to describe the ragged appearance of the rubbish man, or *basurero*, who carries a bag into which he separates 'cabbage, lettuce and cauliflower leaves, pieces of bread and bundles of straw' that he will use to feed his donkey and follows one of the 'seventy rubbish carts which leave Montevideo daily' to its destination at Montevideo's first recorded dump, overlooking the Río de la Plata in the Buceo district and adjacent to the cemetery (Carrasco 2006 [1883]: 36). 'On arrival at the corner', he writes, 'what horror! I found myself in the kingdom of filth, vast, stinking, with mountains of waste and abysses of junk, over which an atmosphere of sour vapours floated, trembling in the light of the sun with dizzying reverberations' (2006 [1883]: 37).

Carrasco continues detailing the materials he encounters: 'here a pile of jars, particularly those of Oriental Tonic, the bombastic hair regenerator from Lanman and Kemp; there a pyramid of bottles; further a stock of broken glass' (2006 [1883]: 38). He catalogues pieces of 'bronze, copper and lead; latches and knockers, lamp tubes, broken gas contraptions, taps, bits of pipe and a thousand other knick-knacks' (2006 [1883]: 38). Set apart is the iron, consisting of 'keys, nails, screws, old bolts, and a hundred other trifles that evade classification' (2006 [1883]: 39). Then there is the zinc and tinplate, which includes: 'pieces of roofing, jars of conserve, tins of oil, pots of paint and varnish, and every other type of fabricated can, all dented, squashed and pierced' (2006 [1883]: 39).

In Carrasco's account, we find organic food waste, a range of metals, paper, glass, bones, and rags. His trip from the household to the landfill is not a literary sleight of hand which elides the industrial and commercial wastes produced in Montevideo: the materials he encounters at Buceo have indeed mostly passed through the household. Industry produced wastes or by-products, but these were not policed or collected by the municipality at the factory gates. Until the 1970s, the Laboratorio's director Joana told me, solid and liquid industrial wastes were often mixed and pumped into rivers and in later years the sewage system, from where they eventually ended up at sea.

If there was a lack of municipal concern for solid and liquid factory wastes at that time, the same did not hold for the gases or 'miasmas' emanating from industry. In mid-nineteenth-century Montevideo, 'doctors, police, press, and citizens shared a deeply rooted belief that unhealthy air generated and transmitted illnesses' (Alpini and Delfino 2016: 383). The central task of the Public Hygiene Council (Junta de Higiene Pública), when it was created in 1836 following a severe outbreak of scarlet fever, was to propose public health measures that would prevent air-borne diseases (2016: 383). The council advised citizens not to throw air-corrupting materials onto the streets, while businesses suspected of being 'the origin of emanations degrading the constitution of the atmosphere' were required to relocate beyond the old walls of the city (2016: 383). Slaughterhouses, brick kilns, soap and candle factories were targeted by decree in 1836, to which were added, in 1868, those producing starch, leather, fireworks, and animal fat (Fernández y Medina 1904).

During the First World War, Uruguay sided with the Allies, while during the Second World War it remained neutral; in both it supplied products such as meat, wool and leather to the warring parties, leading to a period of economic prosperity and a rising population that reached over 2 million in the 1950s. Already, towards the end of the nineteenth century, Uruguayan politicians, conscious of the structural disadvantages of importing most goods from Europe, had sought to impose measures to protect emerging national industries, such as textiles, sugar, and meat processing (Fernández y Medina 1904: 45). Meat in particular benefited from foreign investment and the advent of refrigerator technology that allowed it to cross the Atlantic in large ships – meat-packers in Uruguay are known as *frigoríficos* or 'fridges' (see Edgerton 2008). Taxes were imposed on imported products that could be produced in Uruguay, and duty exemptions were put in place for machinery and parts needed by these industries. Such policies would eventually become known by the 1930s as import-substitution-industrialisation. In other Latin American countries these were specifically tied to centre-left governments – in Uruguay this was less the case, but it is true that protectionist measures would be loosened by the dictatorship (1973–85).

It is widely considered that Uruguay's golden period ended in the early 1960s: some describe a continuous period of growth from the beginning of the twentieth century until then, while others prefer a more accurate stop-start chronology that accounts for dips such as that occasioned by the Great Depression. In 1910, Uruguay was positioned fifteenth in world GDP per capita, the second most prosperous economy in Latin America, behind only its neighbour Argentina (Roman 2016: 22). It was in the 1920s that Uruguay's nickname of the 'Switzerland of the Americas' was coined by a North American journalist, a sobriquet endlessly promoted by subsequent Uruguayan governments. The name reflected not only the standard of living but also the predominantly white European nature of its population (in part due to an effective extermination of original inhabitants), its relative political stability, and, in some accounts, its strong banking sector. In the decade following the Second World War, not only did per capita income increase by more than 50 per cent, it did so as part of an economic model that reduced levels of inequality (Bértola 2005: 28) through policies such as the establishment of collective bargaining (*consejos salariales*).

In the first half of the 1950s, Uruguay's growth was particularly exceptional, with per capita income reaching 90 per cent of that of the richest countries of Western Europe (Bertoni 2011: 90). This period can arguably be described as Uruguay's first period of mass consumption, as the general population was able to access products – such as automobiles and white goods – that until then had been reserved for a privileged elite, with consequences for the generation of waste. As Bertoni (2011) has shown, this was also a period when energy use was 'residentialised' through the electrification of homes, subsidised electricity prices, and higher use of energy at a domestic level, both through electro-domestic goods and the petrol that fuelled new cars. These levels of prosperity are worth bearing in mind as we discuss Uruguay's relationship to European modernity later in this chapter, for as the historian David Edgerton (2018) has argued, 'a southern Spaniard or Italian would have multiplied her income many times over by moving to the River Plate in the first half of the twentieth century', when 'the stuff of modernity was everywhere in Uruguay in a way it was not in Basilicata [Italy]'.

The first comprehensive report detailing the composition of MSW appeared in the late 1950s. This was divided into organic material; paper and cardboard; dust and ashes; tins and metals; bones, meat, and leather; glass and paving stones; and, in much smaller quantities, plastics and rubber (Bonino 1958). Municipal decree No.14.001 (1967) is the first to put forth a definition of household waste, dividing it into 'waste (*desperdicio*) from food and domestic consumption; wrapping and paper from industrial and commercial establishments ... ash and remains (*restos*) from individual heating; pavement sweepings; rubble from small repairs or plant matter from pruning; dead animals; and ashes from the cremation of any of the above'. This decree still holds to this day, with very minor amendments. But when I began fieldwork at the Laboratorio, I encountered a bewildering set of parallel sub-classifications that highlighted an increasingly detailed specification of discarded materials and a more expansive national base of industrial and commercial production. The Laboratorio's somewhat eclectic categorisation was not the only contemporary source of knowledge on the composition of waste. Municipal landfill workers checked the weight of materials that entered, and performed a cursory visual scan to ascertain whether loads roughly corresponded to their paperwork. Landfill waste-pickers, however, were the only ones who truly examined contents in any detail. Rather than a single definitive account, we thus find a set of partial, situated perspectives on the composition of Montevideo's waste. Of these, the Laboratorio's figures may be the most comprehensive and detailed, but in themselves they do not tell us why materials are considered waste in the first place.

For director Joana, it is the act of discarding that converts materials into municipal waste. 'Everything that you have to get rid of (*desprenderse*), that the generator has to get rid of, or is obliged to get rid of when a product doesn't meet a requirement, is waste ... every time that you have to dispose of something, from a control perspective, it is waste for us', she explained. This definition is not based on the composition of waste – whether it might be hazardous or could be safely consumed – absolutely anything that one discards is 'translated' by the municipality's regulatory and classificatory framework into a waste object. This fits with the UK government Department for Environment, Food and Rural Affairs' legal definition of waste, which states that 'a substance or object becomes waste when it is discarded' (DEFRA 2012) and the 2008 European Union (EU) Waste Framework Directive that defines waste as 'any substance or object which the holder discards or is required to discard' (EC DG Environment 2012).

Yet there are other competing definitions, such as that proposed by Michael Thompson in *Rubbish Theory* (2017 [1979]). Thompson starts out with two categories of goods common to economics: transient and durable goods. A classic example of a transient good is a car, which decreases in value from the moment that it is purchased until it is reduced to scrap, while durables, such as certain antiques, 'increase in value over time and have (ideally) infinite life-spans' (2017 [1979]: 25). Thompson's initial question is how an object can cross over from one category to the other, as they do in the case of vintage cars, re-valued pieces of furniture, or works of art. The question precipitates a new third category for goods that are neither decreasing nor increasing in value but are of no value at all: rubbish. Yet Thompson's rubbish category is in fact mostly restricted to a class of objects that still have owners, have depreciated in value, but have not been dumped in landfills or incinerated. Consequently, this constitutes a restricted category of rubbish, excluding what most would think of as rubbish or indeed the discarding practice that I regard as the key moment when objects pass into a waste category. Thompson also has little to say about infrastructures of waste, rubbish, and recycling, a concern which drives this chapter and the book as a whole.

Let us turn then to the *Oxford English Dictionary* (*OED* 2020a) definition of waste then as 'unusable material, substances, or by-products'. I would argue that rather than acting as a technocratic manager of homogeneous substances – wastes – the state often plays an active role in transforming 'waste-as-discard' into 'waste-as-unusable material'. Such was the case when the Intendencia introduced bright orange Mercedes-Benz 'Kuka' dump trucks in the 1980s. Rather than compact waste, the vehicles shredded it. 'Material arrived like a fertiliser or something because the trucks mixed the waste so much that it broke everything down', I was told by Sergio, a clasificador who worked at the landfill at the time. 'It left you some cloth, some big cardboard, some plastic but nothing else … the goods [*mercaderia*] were filthy … it tossed the rubbish together and got everything dirty.' One waste-picker told a newspaper in 1980 that the new trucks were 'limiting the possibility of our subsistence … with the old system we always found elements like plastic, paper, [metals] that after being cleaned we returned to the country as raw material useful for the recycling industry' (Molina 1980). The unintended consequence of this technological development was that clasificadores moved from the landfill back into the central areas of the city to reach discarded materials before they had been shredded. 'All because we bought those damn trucks', lamented the chief municipal sanitary engineer, who preferred having clasificadores confined to the landfill, if they had to be anywhere.

By the time I began working at the Laboratorio in 2014, the origin of materials that entered the municipal landfill had expanded beyond the household to include commercial and non-hazardous industrial waste. Unlike households, businesses were legally obliged to ensure that their wastes met certain criteria in order to be allowed into Felipe Cardoso. They could contain only minimal levels of cadmium, lead, and other heavy metals, and had to be of a solid, rather than liquid consistency. Beyond this, the Intendencia also stipulated that waste had to be rendered 'unusable' (*inutilizado*), meaning broken down or purposefully contaminated. Food had to be shredded and mixed with sodium chloride and sawdust, while other products were dismembered. This stipulation led to a boom in companies whose job it was to neutralise materials so as to render them useless.

I became friends with Fernanda and Homero, jovial, middle-aged, owners of one such company, Stericyclo. When I visited their plant, I encountered hundreds of cartons of a soya milk and fruit juice blend waiting to be squashed and mixed with sawdust; computers being dismantled for their component parts; and cosmetics blended with lime to render them non-toxic and unusable. As part of their social enterprising ethos, the couple employed some former waste-pickers. There was a certain irony in employing workers with a background in waste recovery to destroy discards, however. Some of the materials that they processed were in need of neutralisation from a socio-environmental perspective, such as the phosphorous lightbulbs that the company packed into barrels and sent to Chile for treatment. But the cartons of soya and fruit drink were also 'dangerous' in their own way, Fernanda told me, because they were within their sell-by date and thus 'at risk from theft' by employees. Different conceptualisations of waste and risk and differentially situated 'ways of knowing' (Harris 2007) materials were employed here, and a comparison between them can help bring more clearly into focus the difference between waste-as-discard and waste-as-unusable material.

As I observed from work around the landfill, the clasificador assessment of whether a waste material could be recovered was fundamentally empirical, often involving a sequential use of different sensorial 'tests'. A clasificador who found a bottle of juice would first check to see if the colour inside matched that of the product – they would not drink a liquid from a Coca-Cola bottle that was anything other than black, for example. Then, they would often uncap the bottle and smell the substance inside. I remember bottles of Gatorade that appeared at the landfill: these seemed to be unopened but the colour of the soft drink was close enough to urine to necessitate a quick sniff, just to be on the safe side. Finally, in the case of foodstuffs, the product would be tasted. I recall a bag of biscuits that looked and smelled fine but when I bit into one, I felt the distinct taste of petroleum on my tongue. Where materials passed these tests, clasificadores would declare them '*sano*', meaning healthy, intact, usable or in the case of foodstuffs, good to eat.

While clasificadores would also take into account available information on the provenance of waste, the sensory tests were important if bound by obvious limitations in their ability to detect microbial dangers. In some instances, the Laboratorio and Stericyclo carried out a more detailed chemical and scientific analysis of materials to determine whether or not they were safe to be landfilled with household waste. At first glance then, one could assert that clasificadores had a sense-based epistemological engagement with materials, while waste professionals had a more scientific approach. Yet the nature of the enquiry was also different. Clasificadores checked to ascertain whether or not things could be safely consumed, whereas waste processors tested whether or not they could be safely landfilled. From a municipal perspective, materials had already been classified as waste by the very act of discarding, and so rather than verifying if they were 'usable' (*sano*), it was in many cases Fernanda and Homero's legal and contractual obligation to render them 'unusable'.

What appears from my and clasificadores' perspectives as the wanton contamination and destruction of perfectly good foodstuffs nevertheless had an underlying logic. Debates about food waste have been raging in the UK and Europe for several years, after research estimated that around 90 million tonnes of food are wasted in the EU every year (Evans et al. 2013: 18). EU legislation initially hindered, and latterly has attempted to facilitate, the donation of surplus food. Reasons for companies not donating surplus food stock include concerns about prosecution in case of food poisoning, bureaucratic processes, high costs compared with alternatives, concern about complying with food safety legislation, protection of brand image, and lack of financial incentive (O'Connor and Gheoldus 2014). Some countries have attempted to tackle these issues: Italy and the United States have passed so-called 'Good Samaritan Laws' that protect companies donating food from prosecution, France uses fiscal policy to make it more expensive to send food to anaerobic digestion than to donate it, and Belgium has facilitated contact between potential donors and food banks (O'Connor and Gheoldus 2014). Some of these practices were incorporated into the EU's Circular Economy Action Plan and its food donation guidelines, which advise member states that if food waste cannot be avoided, then the next best option is the redistribution of food for human consumption (European Commission 2017).

The noise about food waste in Europe could be heard in Uruguay, where in 2016 member of parliament (*diputado*) Adrián Peña, now the first Uruguayan Minister for the Environment, citing the example of France, attempted to introduce legislation forcing food producers or distributors to donate any surplus food that could still be consumed. The issue made it onto the national media agenda, alongside the social organisations that received and distributed donated food, such as Banco de Alimentos del

Uruguay (Food Bank Uruguay) and Plato Lleno (Full Plate). These organisations continue to function but the law stalled and a 2017 report found that over 1 million tonnes of food were wasted in Uruguay in 2016, including over 90,000 tonnes (8 per cent) at the distribution stage (Lema et al. 2017). These numbers, including the figure – regularly cited in in the Uruguayan press – that Montevideo's largest fresh food market throws away over 2 tonnes of food daily (de los Santos 2016), do not discount the quantities that clasificadores access informally (we all availed ourselves of market produce at the dump), just as European figures do not account for the food recovered by dumpster divers.

A concern with the recovery of value from waste in the UK constitutes the resurfacing of an old problem rather than the emergence of a new one (Alexander and Reno 2012; Evans et al. 2013: 16), and social critique of food wasting has a particular genealogy in Uruguay. When he drafted the proposed law, *diputado* Peña cited the example of the 'immigrants who constituted Uruguay', the hungry European emigrating masses of the late nineteenth and twentieth century who had known how to conserve and value food (Cambadu 2016). This is the Uruguayan equivalent of the comparison made in the UK to an older wartime generation who were thrifty, managed on rationed food, and grew their own. The narrative ignores the story and values of another often-hungry population, Afro-Uruguayans descended from slaves who form a significant part of the clasificador population. It also glosses over other historical moments when a propensity to waste food was critiqued in Uruguay, such as the European traveller's horror at whole animals left to rot in Uruguay's eighteenth-century low-population, cattle-abundant economy. Echoing the language of John Locke, José Pedro Varela, the nineteenth-century founder of Uruguay's public education system, also critiqued the importation of wine, raisins and dried figs from Europe when Uruguayans 'let figs and grapes rot on the vine because we don't want or know how to use them' (in Faraone 1986 [1974]: 47–8). As he introduced his proposed bill, the focus of Peña's critique had shifted from avoidable 'food loss' at the production stage (which still accounts for the lion's share of food waste in Uruguay and elsewhere) onto the food thrown into supermarket bins (see Gille 2013 for a discussion of the division between 'food loss' and 'food waste').

Both the owners of Stericyclo and the Laboratorio staff had highly ambivalent feelings about the part they played in the 'creative destruction' (Schumpeter 1994 [1942]) of the growing environmental services sector, sentiments that can be noted in the following quote from my interview with Joana:

> For us, the best would be to decrease the amount that enters Felipe Cardoso or the amount that is destroyed. Recycling and re-use are always preferable alternatives. The problem is that we have norms that don't allow for things to be recycled or re-used. There are cases of fake things (*cosas truchas*), brands that are not real. Things come from China, like trainers that say they are Reebok and are not, and one must destroy them (*hay que destruirlos*). For us, it would be better if these things were distributed among children in care, people who are in prison, I don't know. But then there is the issue of 'No, if you give it to them, they might sell it, commercialise it, and it would return to the market', all those issues which are outside our remit (*escapan a nosotros*).

In the above passage, Joana moves between the Spanish first-person plural 'we' (*nosotros*) and the impersonal 'one' (*hay que*) to navigate the moral economy of waste disposal/creation. She repeats that *para nosotros* ('for us') – gesturing to the Laboratorio's employees – materials should in the first instance be recycled and re-used, and 'fake' brands distributed to vulnerable groups. It is impersonal 'norms' and issues beyond the Laboratorio's control that mean that *one* 'must destroy things' (*hay que destruirlos*), preventing their ethical redistribution. Yet Laboratorio staff play an active role in drawing up legal definitions of waste and in policing that things are correctly destroyed. 'If it can't be sold, it can't be donated', Joana clarified.

The kind of material transformations that Fernanda and Homero engaged in demonstrate the diversity of practices that currently take place in Uruguay under the banner of 'environmental services'. On the one hand, they employed two technicians to disassemble computer parts, and Homero proudly told me that 98 per cent of these could re-enter productive processes and avoid landfill. On the other hand, they also destroyed crates of foodstuffs like the soya and fruit drink that might have been consumed. I was uncomfortable with this practice, as I knew that had the cartons not passed through municipal oversight and the hands of workers then they might have arrived at Felipe Cardoso intact and, after being tested by my clasificador neighbours, declared *sano* and consumed.

Indeed, I was reminded of the attempts to police waste back in St Andrews. At first, the supermarket bins where we recovered discarded food had been unguarded, and we had been unconcerned about revealing our identities. As our presence became felt, however, the supermarket first installed glaring nightlights, not dissimilar to those that I would later encounter at Felipe Cardoso. When the lights failed to deter us, security cameras were installed, leading to the need for us to cover our faces. Since the local police already had us in their sights for our anti-war and environmental activism, we couldn't risk arrest. Finally, locks began to be placed on the bins, and when we cut through those, larger locks would appear until finally one of us came up with the ingenious idea of gluing the lock closed. Only after this final ruse did the supermarket appear to give up, and give us free rein over the food products that were, in the words of my Uruguayan interlocutors, perfectly *sano*. What was insane was their being thrown away on such a mass scale.

Fernanda told me that she often received materials that were 'perfectly reusable' but had to be destroyed because of 'legal factors' and 'accounting/ inventory practices'. For such materials to avoid destruction, she noted, 'the law would have to change'. Like Joana, the couple also found the destruction of value problematic. Homero said that for each of their workers, the treatment of a particular material might act as a 'trigger' for affective and ethical concerns. For some workers, the agentive discards might be expensive make-up sets or bottles of juice, while for Homero it was the destruction of functioning computers that he found most difficult. As we can see then, the Montevidean municipality and environmental management companies do not merely manage waste that is somehow already 'out there'. Rather, through decree and infrastructural provision, they play an ethically ambiguous role in transforming 'waste-as-discard' into 'waste-as-unusable material'. The materials are not 'unusable' in the last instance – a use might be found for shredded and disassembled things – but they are unusable for most clasificadores and thus tend to be beyond recovery at Felipe Cardoso.

Hygienic enclosure and infrastructural modernity

When waste was born as an object of municipal control, so too was an infrastructure designed to contain, transport, process, and eliminate it. In the Uruguayan case, enclosure devices have played a constitutive role in the definition of waste. The coal dust, packaging, rubble, food waste, and prunings that an early twentieth-century Montevidean home co-produced were heterogeneous materials brought together in the category of household waste by municipal decree but also by emplacement in specific containers. According to a 1937 decree, the classification of material as 'household waste' was contingent upon its ability to 'fit within one or several of the containers normally used'.[2] Non-household waste, meanwhile, was defined negatively as that which 'exceeds the proportions indicated'.[3] A 1959 decree excluded from the category of household waste materials that did not fit into 50m² containers,[4] while another stipulated that waste should be 'handed over to the authorities in buckets with particular characteristics specified in the accompanying regulations', where 'the content cannot by any measure exceed the recipient or fall on the floor'.[5] For a material to be classified as MSW then, it was not that it must somehow be 'out of place' (Douglas 1966) but rather it should fit in a particular place: the municipal waste container.

The penchant for detail in specifying infrastructures of waste containment has been matched historically by the Intendencia's willingness to spend large sums of money importing technologies of enclosure. Just as there always seems to be money available for the machinery of war, so too has it been found for the artillery deployed against the enemies of hygiene. In Uruguay, the two spheres often overlapped and we should perhaps not be surprised – Adam Smith (1896: 154) did, after all, list cleanliness, 'the proper way of carrying dirt from the streets', alongside security and ensuring plenty as constituting the basic functions of the state. One famous and controversial campaign, Operation Clean-Up (*Operativo Limpieza*) took place in 1967–8 under the direction of Colonel Manuel Díaz Ciblis. Contemporary opinion pieces complained about the exorbitant sums that were being extorted from the population for the operation, without visible result. But the municipal head of Engineering and Works argued that it was necessary in order to 'defend the city' from the 'filthy state, lack of hygiene, insects, rodents and illnesses' generated by wastes (*El Diario* 1968).

The emphasis on keeping waste contained while it is temporarily held on the city streets has, as in the case of Bonino's 1958 report, often been framed in a hygieno-aesthetic register bordering on the apocalyptic. Failure to attend to the elimination of waste led, in the municipal engineer's view, to 'immediate hygienic and aesthetic chaos' (Bonino 1958: 17). 'Rubbish will cover the streets of the city' (*La Mañana* 1971); 'Deficient hygiene in the city' (*La Plata* 1966); 'Montevideo: anti-hygienic city' (*BP* 1966), screamed newspaper headlines accompanied by photos of the offending matter overflowing onto street corners. In such accounts, Montevideo appears as a 'vast rubbish tip' (*El País* 1971), 'a septic hotspot' (*El*

Día 1970) or even 'the city with most rubbish in public space, per capita and per square metre' (*El Día* 1970). This was then principally a 'hygienic enclosure', but one which also involved questions of ownership, aesthetics, value, and modernity, as properly enclosed materials became municipal property. Early Montevidean waste ordinances, sparing in their description of MSW's composition, all featured articles prohibiting the extraction of materials once they entered containers and detailing the corresponding sanctions. The 1937 decree, which stipulated that 'from the moment in which domestic waste and street sweepings are left out to be collected by the Public Hygiene Authorities, the extraction of any materials from containers is prohibited', remained in force until 1991.

When a new fleet of trucks was purchased from Argentina in 1971, these were lined up in front of the Intendencia alongside an equestrian statue of Uruguay's founding father, General José Gervasio Artigas, giving them the appearance of tanks accompanying him into battle (*Vea* 1971). When 3,200 standardised municipal waste containers were introduced in Montevideo in 2003 at the cost (together with corresponding trucks) of over US$5 million, the newspaper *El País* (2003) described their arrival as an 'invasion'. The containerisation of Montevideo was intimately connected to the privatisation of part of the collection service, since it was the private city centre concessionary CAP that had pushed for the replacement of informal 'baskets' with containers. In 2014, the public purse again forked out for 'anti-vandal' containers purchased for the Ley de Envases and shiny new US$500,000 trucks fitted with Italian technology that were to be operated by CAP. 'The new containers also have a more modern design', a CAP coordinator indicated to me, 'which allowed us to provide a change in the aesthetics.' The containers had been adapted especially so as to prohibit the unlicensed extraction of waste, an indication of how such devices, though originating in the Global North and embedded in unequal transnational economic relations, can be reprogrammed to materialise local ethico-political projects (Von Schnitzler 2013).

The concept of 'infrastructural modernity' (Collier 2011) is useful for scoping out the horizons for Uruguay's technologies and priorities in waste management. Early works in the study of infrastructure, such as Paul Edwards' (2002) 'Infrastructure and modernity', have noted how bound up modern societies are with infrastructures understood as information systems, standards, and built environments of connection and communication such as train tracks, roads, and telephone lines. To this foundation Collier (2011) adds the point that infrastructures are political: they are programmed to meet certain objectives, enable certain flows, and restrict others. Infrastructural modernity is thus plural rather than singular: it comes in many forms, and to identify them we must study 'how infrastructure is mobilized as a political technology' and the goals and forms of reasoning involved in its development (Collier 2011: 205). Waste collection, in common with other infrastructures, ideally should function as an 'invisible, smooth-functioning background' (Edwards 2002: 188), noticed only when something goes wrong – 'when the server is down, the bridge washes out, there is a power blackout' (Bowker and Leigh Star 1999: 35). Or, in the case of waste, when containment is breached. This is at least the case in the cities of the Global North, from where the early theorists of infrastructure drew their examples.[6] More recently Dalakoglou and Kallianos (2014) have argued that for southern European cities creaking under imposed austerity, 'when it comes to ... waste infrastructure and its flows, disorder and arrhythmia are part of the normal infrastructural patterns for the people who have direct experience of it' (2014: 531). Further, the dimensions of modernity that include increased purchasing power, consumerism, and disposable packaging can in fact overwhelm the hygienic management of cities of the Global South that lack the infrastructure to deal with the volumes of waste generated as a consequence (Moore 2009; Stamatopoulou-Robbins 2020).

Discussions of modernity have a long history in Uruguay, but these have not generally been linked to waste management. Focusing on the connection here can help us to understand why waste-pickers have generally been seen as part of the problem rather than the solution. In waste and hygiene, as in other areas, Uruguayan horizons of modernity have always been primarily European, from early containers based on French designs, through English-built incinerators, to present-day Italian waste trucks and waste-to-energy models. Indeed, the challenges that containerisation posed for waste-pickers in Montevideo mirrored those which affected Parisian chiffoniers a century earlier, when a certain *Monsieur Poubelle* pioneered the waste receptacle (Barles 2005). When I asked a municipal waste official which cities she viewed as exemplary in their waste management, she cited cases from the US, Sweden and Spain, which had 'clean incinerators even in the middle of cities'. She dismissed Latin America as 'very backward', disregarding the important experiments in 'inclusive recycling' in cities like Bogotá and Porto Alegre (Red-LACRE 2017) in favour of technological solutions designed for cities without waste-pickers. Landfilling, even of the sanitary kind has, as Stamatopoulou-Robbins (2020:48) describes in her ethnography of Palestinian waste practices and management under the Israeli occupation, fallen behind as a marker of modernity for contemporary waste managers, preferable to dumping, certainly, but lacking the technological sophistication of new forms of waste-to-energy or gasification and pyrolysis.

It was not only physical technologies that were imported from Europe but also the very idea of what a municipal waste management infrastructure should look like. We thus find the influence of early French and English hygienists, and the 'infrastructural ideal' of state-led universal infrastructure that prevailed in many parts of the world from the mid-nineteenth century well into the twentieth. Graham and Marvin (2001) argue that a splintering of municipal provision characterised a neoliberal turn in much of the West from the 1970s onwards. Until the 1980s in Britain, for example, most waste collection and treatment services were owned and operated directly by local governments. Under Margaret Thatcher's Conservative government, 'contracting out' was one of three policies used to shift the balance of public and private ownership in the UK, the other two being the wholesale privatisation of industries and the transfer of property ownership through council house sales (Davies 2007). Yet it was not until 1988 that Compulsory Competitive Tendering (CCT) was introduced for refuse collection, through the Local Government Act. A similar directive for waste disposal operations came with the 1990 Environmental Protection Act (EPA), which forced local authorities to create arm's-length local authority waste disposal companies and to compete for contracts with private sector companies. With the arrival in power of the New Labour government in 1997, the compulsory element of CCT was removed but a strong focus on competition was retained, and the current UK market in both collection and disposal is characterised by a concentration of large multinational companies like Biffa, Veolia, and Suez at the expense of smaller regional firms and public sector provision (Davies 2007).

Such a neoliberal turn also manifested itself in Montevideo, with the privatisation of city centre waste collection in 1995. Further splintering was, however, resisted. Attempts to privatise the Felipe Cardoso landfill in 2009 were blocked, and waste collection largely remains in the hands of the Intendencia, not least because the centre-left Frente Amplio has controlled Montevideo's local government since 1990. The head of waste management during my fieldwork period hailed from the Uruguayan Communist Party. Yet it is debatable whether the conservation of the status quo in Montevidean waste management is a function of ideological conviction or simply of the municipal government defending its administrative and operational territory. For example, in 2014 the national Office for Planning and Budgets (OPP), together with a group representing Uruguayan state governors, proposed a new waste-to-energy plant that would incinerate waste from Montevideo and its surrounding states. The Intendencia strongly objected, not because this would have been catastrophic for clasificadores, but because it would have meant a huge loss of municipal influence and control. In the end, the scheme floundered largely due to the lack of financial viability and high projected electricity costs. Such an outcome is reminiscent of another centralised, technological waste management scheme proposed in Cusco and killed off by the 'objections of local municipalities' (Harvey 2017: 683). All the same, trade union calls to re-municipalise city centre waste collection have thus far fallen on deaf ears. Consecutive Frente Amplio governments have consistently renewed CAP's concession with the current (2019–22) contract worth US$12 million (López Reilly 2018), a sum predicated on the frequency of collection and additional street-sweeping activities rather than overall tonnage (see Figure 4).

I would suggest that Montevidean waste management has been characterised by a hybrid condition. The municipal government has been able to afford imported technologies of containment, and elites are largely Europhiles. At the same time, infrastructures of waste containment and disposal have broken down with alarming frequency, putting Montevideo's status as a modern city at risk. 'The Switzerland [*Suiza*] of the Americas or the dirtiest [*Sucia*] country of the Americas?' (*Epoca* 1966), punned one headline, referring to Uruguay's prized epithet: clearly it couldn't be both. More recent reports feature a similar aesthetic of overflowing bins, and the literary conceit of punning on Uruguay's previously flattering nicknames (*El Observador* 2014b; *El País* 2016). We might contrast the regular breakdowns in Montevidean waste management with waste infrastructure in Britain, where the presence of waste piled high in the streets during the 'Winter of Discontent' was a singular, epoch-defining moment that arguably helped to bring Margaret Thatcher to power amid industrial strife (Martín López 2014), leading ultimately to a neo-imperial push to reassert British presence in Latin America, particularly through the Falklands War with Argentina. Interestingly, Uruguay's port infrastructure is the latest to be drawn into this ongoing dispute, as the country has refused to let ships sailing to the Falklands or flying the Falklands flag dock in its harbours (UPI 2011). Even more recently, Argentina has been pushing Uruguay to adopt a similar policy for Falklands aircraft (Mercosur Press 2020).

At different times, municipal workers, technological breakdown, disorderly citizens, and managerial ineptitude have all been blamed for the overflowing waste on Montevideo's streets. But as Stamatopoulou-Robbins suggests, 'waste's murky indexicality can invite discursive displacement of its burdens onto single actors' (2020: 8) and in Uruguay it is waste-pickers who have most regularly been singled out as the principal agents of disorder. In moralised and politicised universes, already unpopular groups are often blamed for new problems or disasters (Douglas 1992: 5). Thus in 1985, Montevidean mayor Aquiles Lanza affirmed that waste collection was his administration's gravest problem and that 'rummagers' (*hurgadores*) were partly responsible for spreading waste on street corners (*La Mañana* 1985). In 1989, the municipal director of public hygiene complained that the two biggest contributing factors to the presence of rubbish in the streets were 'rummagers' and the absenteeism of municipal workers (*El País* 1989). In 2016, the Intendencia's outgoing head of waste collection and street-sweeping declared that Montevideo's biggest problem was the 'illegal classification of waste' (*Telemundo* 2016).

Clasificadores are so often scapegoated, I would argue, because they challenge the 'infrastructural modernity' of Montevidean waste management. First, they dispute a municipal monopoly whereby the local state either operates infrastructure itself or holds the power to grant concessions. Second, with their uncovered horse-and-cart collection and suspected dispersal of rubbish, they challenge long-established technological norms and hygienic prescriptions. For many municipal bureaucrats and city planners, waste-pickers represent a sort of 'heterocronia' (Fernández 2010: 2) existing, as one newspaper put it, 'three blocks from the centre but centuries away' (*Últimas Noticias* 1986).

Eliminating surplus material and populations

Thus far, we have tracked the creation, classification, and (attempted) containment of Montevideo's waste. The imported municipal truck now arrives to empty street-level containers, perhaps scooping up the debris that invariably surrounds them. What happens next? Over the last 150 years, Montevidean waste been taken to different locations for elimination – sites that have also adapted to conform with hegemonic ideas of hygienic and infrastructural modernity. The late nineteenth-century Buceo landfill economy described by Daniel Muñoz was buried by Montevideo's early twentieth-century 'hygienic modernists' (Rogaski 2004). As part of a 1914 committee that sought to solve the city's waste 'problem', the architect Juan M. Aubriot railed against the insalubrities of the Buceo dump, where germs had replaced miasmatic 'sour vapours' as the source of disease (Alfaro 1971).[7] For most of the nineteenth century, city officials had focused on industrial production and cemeteries as miasmatic threats to public health. At the beginning of the twentieth century, they turned their attention to landfills.

The state concessionary Meneses dumped its last cartload of rubbish at Buceo in 1915 and the country's first incinerator, Usina 1, was built further down the coast before the end of the decade (*El País* 1967). Henceforth, tourists arriving on boats or frequenting beaches would be met with a different kind of dust: clean black chimney smoke to which Montevideo's heterogeneous waste was reduced, inaugurating the waste technology that would reign supreme for the next fifty years. The fact that Usina 1 was situated closer to the centre of Montevideo than the Buceo dump suggests that officials believed it to be completely effective in eliminating the risks of urban waste. 'At the beginning of the [twentieth] century, Montevideo was very advanced in waste treatment compared with the rest of Latin American countries', the municipal engineer told me proudly. 'We incinerated household waste and … had three Usinas and [each] had its own chambers for burning waste.' He boasted of the use of English technology, an early indication that Montevideo's conceptions of modernity were primarily imported from Europe. The location of the first *usina* (waste plant) also serves as a metaphor for such horizons, sitting as it did on the fringes of the city's landmass, edging out into the Río de la Plata and beyond it the Atlantic (see Figure 3). As the old national saying went: 'Uruguay with its back to Latin America, Montevideo with its back to rest of the country, the city centre with its back to the periphery' (Alfaro 1971).

The desirability of incinerators can be understood if we recognise that sanitation and waste management infrastructures emerged primarily as defences against hygienic and epidemic risk. In England, where the first waste furnaces were designed and patented in 1874, fire was thought to 'permit the perfect destruction of "contagia and virus"' (Clark 2007: 261). Before being imported as indicators of infrastructural modernity in Uruguay, 'destructors' in England were heralded as the progressive alternative to primitive tipping, which was denounced as a 'miserable link with

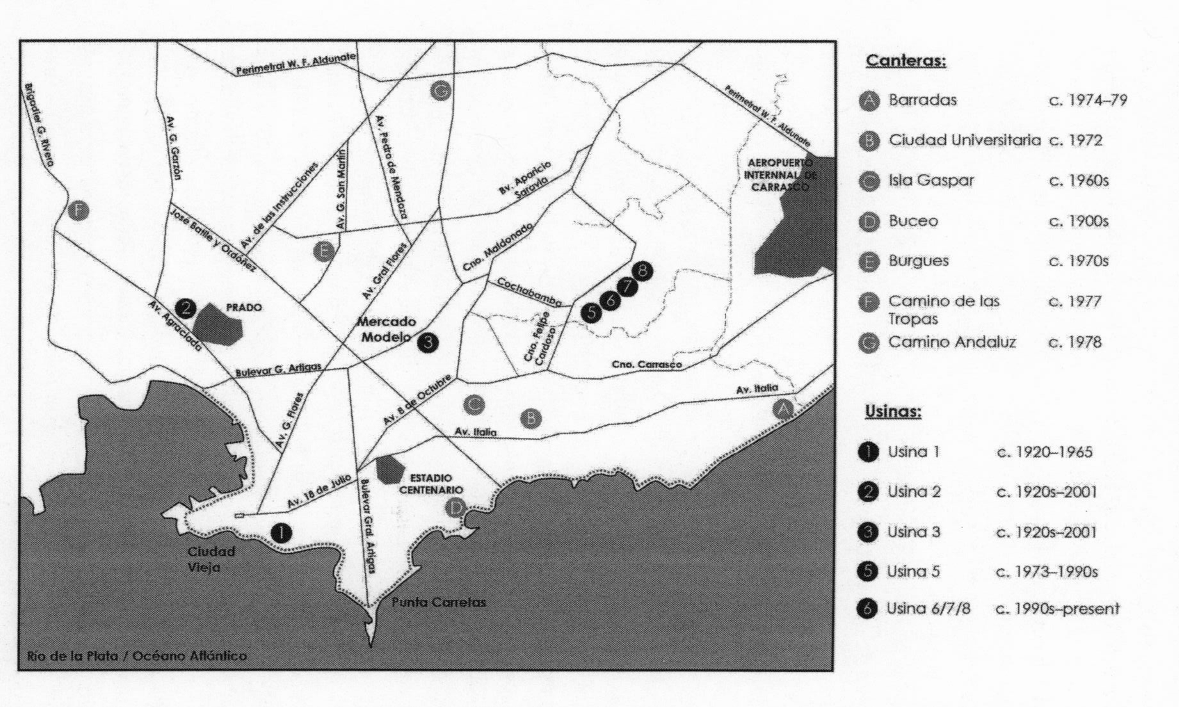

Figure 3 Location and estimated operational dates of Montevideo's dumps (*canteras*) and *usinas* (waste plants)

Source: Map designed by Mary Freedman and Adriana Massidda.

the insanitary past' (Clark 2007: 261).[8] In Montevideo, the first *usina* was followed by the construction of Usina 2 in the west of the city and Usina 3 in the east, where the burning of waste continued for the first half of the twentieth century. The 1937 City Hygiene Act also obliged any residences producing large amounts of waste to 'install incinerating ovens for the elimination of such waste' (Article 23). The burning of waste formed part of an urban landscape dominated by 'chimney dreamers' (Gudynas 2004) who saw in smoky industry the symbols and materialisations of modernity.

In the late 1950s and early 1960s the Uruguayan economy underwent a period of 'stag-flation' (stagnation with high rates of inflation) and social and political divisions deepened. After the coup d'état of 1973 and the establishment of the so-called 'civic-military government', the bargaining power of the Uruguayan working class was severely diminished by the disarticulation of collective bargaining (already suspended in 1968) and the repression of the trade union movement. Unemployment increased steadily if unevenly, from below 9 per cent in 1973 to 14 per cent in 1984, the eve of the democratic transition. The dictatorship also failed to control inflation in any sustained way, inheriting a rate of 77.5 per cent in 1973 and ending with a rate of 66.1 per cent in 1984, with a fluctuation between lows of 20 per cent and highs of over 100 per cent in the interim (Yaffé 2009: 22). With wages stagnant, this meant that real wages were, by the end of the dictatorship, almost half of those of 1973 (Yaffé 2009: 22). As Yaffé surmises then, the dictatorship produced a 'spectacular reduction in workers' acquisitive power', despite the fact that GDP had doubled between 1973 and 1984 (2009: 22). A new, unequal Uruguay had been created that would not be seriously challenged until the centre-left Frente Amplio arrived in power with redistributive policies (so-called 'growth plus distribution') in 2004.

These trends matter for our present discussion because levels of consumption clearly influenced the volume and composition of waste and recyclate generated: the very materials from which a recycling industry and a clasificador livelihood can be sustained. In general, a booming economy tends to produce greater levels of waste. In Uruguay, a recent example of this is the prolonged growth period from 2005 to 2012, during which time the amount of waste that the private concessionary CAP collected in Montevideo's city centre doubled. Nevertheless, techno-economic shifts such as the widespread adoption of plastic for a range of products are not necessarily dependent on periods of boom and bust. According to Faraone (1986 [1974]: 111), from the end of the Second World

War, foreign, and particularly North American capital, contributed to the creation of new industries in Uruguay such as metal, electronics, rubber, and chemicals, including plastics. The Uruguayan Plastics Industry Association (Asociación Uruguaya de Industria de Plástico – AUIP), dates the origins of the plastics industry to the mid-1940s, with the production of Bakelite radios and light switches, followed by polyethylene films, cellulose acetate hairbrushes, and acrylic boards. The first plastics producers' association was created in 1956, and, as the AUIP argues, plastic began to 'substitute and improve the design of products traditionally made with glass, metal, and wood' (AUIP 2020).

As they began to enter Uruguayan waste-streams, plastics undermined what officials called the 'safest sanitation technique' (*El País* 1967), as incinerators started collapsing under the weight of increasing volumes of waste and its changing composition. Usina 1 was closed in 1965, its chimneys dynamited and themselves reduced to rubble (*El Diario* 1969). The other *usinas* deteriorated and the introduction of plastics was held responsible (*El Diario* 1974). The then-municipal director explained that 'when waste is burnt in the *usinas*, the corrosive smoke given off by plastics seriously damages the ovens ... with time, the thick metallic plates and the pipes disintegrate' (*El Diario* 1974). Individual waste incinerators were eventually also banned, amidst complaints of smog and air pollution (*El País* 1980). Plastics were 'biting back' (Tenner 1996), exposing the limits of fantasies and technologies of total elimination.

Incineration began to be replaced with a return to what one newspaper called the 'human and sanitary quagmire of dumping in *canteras*' (*El País* 1967). Yet although the source of waste's perceived risk (miasma, germs) and the method of elimination (incineration, landfill) changed, what remained was a logic that conceptualised waste negatively as a source of risk and elicited technologies designed to eliminate its capacity for sanitary and aesthetic contamination. As a polarised Uruguay slid into urban guerrilla warfare and state terror in the 1960s and 1970s, the treatment of waste was also increasingly discussed in martial terms. It was the 'fight against the dump' (*Acción* 1968), 'the war against filth' (*El País* 1967) and the 'battle against plastic' (*El Diario* 1974) as Montevideo was 'invaded by rubbish' (*El País* 1971).

Clasificadores were often portrayed as enemies, not allies, in these wars. Already in 1965, a group of middle-class neighbours occupied a landfill to protest against 'an incredible invasion of flies, bad smells and on top of these calamities, almost a hundred *cirujas* (waste-pickers) who rummaged in the rubbish and whose customs and manners were at odds with morality' (*Acción* 1968). In 1967, a newspaper report described the *canteras* as 'immense pits where chilling sub-human stories ferment between the bubbles of hate and filth' (*El País* 1967). Those found working there were dehumanised in the press as 'strange beings disguised as men, women and children' (*El País* 1967). The language used to describe these workers facilitated their denigration.[9] In another report in 1980, a bleak photo of grey figures at the *cantera* of Felipe Cardoso was accompanied by a sub-heading which clarified that these were 'also Uruguayans' (Molina 1980: 19). With non-naturalised citizens only minimally present in Uruguay at that time, the question mark seemed to hang less over their nationality than their very humanity. Bowker and Leigh Star (1999:1) may argue that 'to classify is human' but in this context getting one's hands dirty with waste, even to classify it, seemed to index infra-human life.

In neighbours' complaints about the 'activities of so-called *bichicomes* and the dumping of rotting organic matter' (*Acción* 1968), the threads of waste and those who made a living from it were woven in a Gordian knot to be solved by a radical solution: elimination. The military dictatorship, buoyed by the dehumanising language of the press and middle classes, enforced waste law in its severity and beyond. In 1976, the Intendencia placed an advertisement in a mainstream daily, informing 'rummagers' that they could not 'under any circumstances undertake the collection of waste or circulate on public roads' under pain of 'severe sanction' and the confiscation of vehicles (*El Día* 1976). In 1979, the chief inspector of vehicles threatened 'a frontal fight to eradicate ... vagabonds from the Old City' (*El Diario* 1979), while in 1980 an operation was undertaken to 'eradicate rummagers and their carts' with 'brigades traversing the city repressing rummagers day and night' (*El País* 1980). An editorial in one newspaper expressed disbelief at the Intendencia's burning of several hundred carts, which 'had been constructed with sacrifice by whole families, and were their only work instruments, the only tool which helped them to honourably earn their daily bread' (*El Día* 1980). Although this population appeared as surplus to the dictatorship (Bauman 2003), they were integral to the recycling industry, as evidenced by the paper manufacturers, who complained in the press about the effect that the crackdown was having on their production (see Millar 2012).

During the dictatorship, there was thus a dual focus on eliminating both the sanitary risks of polluting things and the perceived political risks of the 'dangerous classes'. The reintroduction of burial applied not only to wastes but also to mangled bodies thrown into mass graves, many still undiscovered today. Zuli was a middle-aged colleague at the Aries recycling plant when I conducted participant observation there in 2014, but she had worked as a clasificadora since moving from rural Rivera with her grandmother as a young girl in the 1970s. In those years, she remembered making her way to an old country house and grounds where the Intendencia operated a landfill for a short period of time. It has since emerged that the site was used as a detention centre for so-called 'subversives', some of whom, it is suspected, were killed and buried there (*La Red 21* 2007). The macabre link between the graveyards of persons and things thus re-emerges, coextensive rather than contiguous this time, both victims of the eliminatory zeal of the fascist generals.

Elimination did not constitute a temporally constrained 'waste regime' (Gille 2007) that disappeared at the end of the dictatorship. Instead, a logic of elimination focusing on the dangerous properties of both waste and *hurgadores* remained 'residual' (Williams 1977) and resurfaced in different guises. Attempts to extend municipal control over waste, and minimise its risks through more closely monitored landfill practices, prevailed. By the time of my fieldwork period of 2014, Felipe Cardoso had been recognised as a 'controlled', if not a fully sanitary landfill, with additional plants built to capture and burn methane and to pump out and treat leachates. Despite Joana's desire to minimise the amount of waste entering the dump, 70 rubbish carts at the beginning of the twentieth century had grown to 700 mechanised trucks dumping over 2,000 tonnes of waste on a daily basis (LKSur et al. 2013: 8).[10] The normative, linear process of waste generation, collection, and disposal remained hegemonic.

Today, most of the goods disposed of are imported, and consist of hybrid and synthetic materials. In recent years, Uruguay has followed many countries in increasing the quantities of consumer and industrial goods imported from China (Bartesaghi and Managa 2012). In 1992, Chinese imports to Uruguay totalled US$18 million, of which 40 per cent were garments and clothing. By 2011, such imports had risen to US$2 billion, of which 33 per cent were electronic machines and devices and only 10 per cent clothing (Bartesaghi and Managa 2012). Toys, rubber products, furniture, plastics, textiles, and leather also featured among the imported goods, most of which eventually make their way to the landfill. Unlike in some 'developed' countries, from which the Montevidean state imported technologies, hygienic norms, and standards of modernity, Uruguayan technocrats have had to account for the 'shadow infrastructure' constituted by waste-pickers. Yet the clasificador families whom we meet in this book are not a class of people who live from whatever another class of consumers throw away, without consuming themselves. Rather, clasificadores both consume new products, including those that are most symbolic of consumer-capitalism, such as the latest trends in children's toys or Nike trainers – *and* those that have been discarded, either by industry or by households. I would suggest that this point holds for global waste-pickers more generally: they rely on mass consumption and disposal but also feed into this system by consuming themselves, even in Uruguay's neoliberal consumerist meccas par excellence, *los shopping* (Di Stefano 2012). Far from being eliminated then, both clasificadores and waste flows have to a large extent increased, and now form an integral part of the importation, production, and disposal of commodities.

Figure 4 Protesting waste-pickers make their way through Montevideo's old town as a worker for the privatised city centre sweeping and waste collection service looks on

Source: Author photo, 9 December 2013.

Shadow infrastructures on the waste commons

Montevideo's local government has only in recent years, and through partnership with the private sector, provided a limited recycling infrastructure for the city's residents. This did not, however, mean that materials were not previously classified or recovered. A widely cited report suggests that around 400 tonnes per day are saved from burial in the municipal dump by the work of clasificadores (OPP et al. 2005: 267), generating a saving for the Intendencia of approximately US$65 per ton. Lucía Fernández (2012) has suggested that up to 52 per cent of Montevideo's waste is recycled by waste-pickers. Indeed, just as waste can be considered the dark shadow of more visible processes of production and consumption (Moore 2009), as far back as the origins of Montevideo's municipal waste management, clasificadores represented a shadow infrastructure concerned with recovering value from the waste commons. Thus, inside the 'kingdom of filth' he described at the nineteenth-century Buceo landfill, Daniel Muñoz found:

> men who, like pigs, root in the rubbish, disputing with them the scraps. Nothing is wasted, everything is classified and collected separately: bones here, rags there, beyond them tins and leathers, everything neatly removed from the rubbish that the city's throws away daily as if useless waste. The leftovers of Montevideo support an industry, a productive industry that provides work to hundreds of arms and feeds numerous families, as well as a thousand succulent and respectable pigs. (Carrasco 2006 [1883]: 37)

As we have seen, the messy but ultimately productive inter-species ecosystem of Buceo proved offensive and dangerous to Uruguayan hygienic modernists and was replaced by the fantasy of total elimination embodied in technologies of incineration. But when this infrastructure began to falter and landfills were re-established, waste-picker families followed and numbers burgeoned. A 1986 local government study (BIRF-IMM 1986) cited by Chabalgoity et al. (2004) provides one of the earliest clasificador estimates. For the year 1978, it gives the number of 800 waste-pickers divided between 600 working at the landfill and 200 horse and carts in the city. The second figure taken up by Chabalgoity et al. (2004) is from the same study, and describes the number of carts for 1986 as having grown to between 2,000 and 3,000. The numbers are slippery – several people might work collecting and separating materials from a single cart – but they appear to capture an inverse relationship, suggested to me by municipal interviewees, between the numbers allowed to work at the landfill and those on the streets.

During a brief period in the 1970s before dumping was centralised at Felipe Cardoso, Montevideo's Intendencia operated short-term dumping grounds at various points of the city: Oncativo in the east, Barradas in the south, La Tablada (Camino de las Tropas) in the west, Camino Andaluz in the north and Boi Merino y Menorca nearer the geographic centre (see Figure 3). As with the Buceo escarpment, there is nothing visible in the seaside gated community of Barradas, located on the fringes of Montevideo, that suggests the tonnes of waste which lie beneath. Yet in the 1970s, this was a landfill where a teenage Zuli turned up with her grandmother after falling on hard times. Their tools consisted of hooks on sticks (*ganchos*) with which they raked through the rubbish, but the women would disguise these in musical instrument cases so as not to betray the stigma of their occupation. '"There go the guitarists!", other passengers would say, and my granny would kill herself laughing', Zuli giggled. 'We went all dressed up as if we were going to [the middle-class seaside neighbourhood of] Pocitos but we were going to the dump! We'd arrive and get changed into old clothes.' 'I liked Barradas because the beach and the River Carrasco were nearby; I especially liked break time at midday when the trucks which brought the goods [*mercadería*] stopped and ... I would go off to explore the forest, the plants, the trees', Zuli reminisced. 'I stayed close to the river because I was used to playing by the stream and woods in Rivera and I was reminded of my village.' 'It was strange because I loved to go to the *cantera*', she continued, 'we gathered [*juntábamos*] bread, bones, glass bottles, cardboard, and metals, leaving paper for others.'

Zuli accompanied her memories of bathing at Barradas with others: picking flowers at Felipe Cardoso, and even scrumping for peaches as she wandered to the dump at Camino de las Tropas. These images contrast sharply with the dystopian malaise of foul smells and creatures found in risk-based municipal and journalistic descriptions of dump sites. For Zuli, recovering materials was a way of avoiding hardship in the city; landfills and their surrounding environments were spaces of refuge, leisure, and even beauty. While municipal trucks dumped *residuos*, Zuli and her colleagues recovered *material* or *mercadería*. Although discarded plastics caused problems for technologies of elimination, clasificadores at the *cantera* were the first to realise their potential value. 'Before, no-one collected the *botellita* [little bottle]', my 60-year-old neighbour and veteran clasificadora Beatriz explained, referring to plastic PET bottles. 'Who first started working the bottles because no-one else did? Me', she smiled proudly.

Clasificadores have largely appeared in this chapter as a group who interfere with the smooth running of municipal infrastructure. But their presence also influenced the planning of it. The obligation to destroy materials – to transform them from waste-as-discard to waste-as-unusable-material – responded not just to a generic hygienic imperative but to the presence of waste-pickers at the landfill. To take the case of another material I witnessed being destroyed, this time at the Intendencia's composting plant, what risk did packets of Phillip Morris cigarettes pose to the operation of the landfill? The risk, which existed as a sort of 'open secret', was that clasificadores at the landfill would either consume them or put them back into commercial circulation. This applied to even the most unlikely items. When she received the phone-call enquiry about disposing of shark meat at Felipe Cardoso, Joana responded that this certainly couldn't go in as it was, because clasificadores might eat it. I don't know whether shark meat would have held much interest for my interlocutors, but then I couldn't work out whether their tale of eating Leo (the much-loved first elephant to be born in Uruguay) after he died and was dumped in Felipe Cardoso, should be taken seriously. In the final weeks of my fieldwork, my neighbour Martín Azucarero also did the rounds selling Australian steaks that had turned up at the landfill, presumably after having been impounded at the port. Joana's fears about illicit shark consumption might well have been justified.

The presence of urban waste-pickers shaped municipal waste infrastructures in other ways too. As part of a large Inter-American Development Bank (IDB)-funded project to improve Montevideo's sanitary system in the early 2000s, recycling stations known as 'green points' (*puntos verdes*) were set up so that clasificadores could leave their discards there instead of clogging up pipes and waterways with them. In 2014, when the Intendencia ordered special sliding-mouth waste containers for the city centre, there was little doubt that these were designed to keep out not 'vandals', as was claimed, but waste-pickers. At Felipe Cardoso itself, the foreman told me that waste was being dumped in such a way so as to build up a great wall of rubbish, behind which waste would continue to be landfilled, and waste-pickers continue to labour without being visible from the street. Waste technologies were sometimes designed to make the work of clasificadores difficult then, and at other times to accommodate their presence.

In the Laboratorio, Joana and her team played their own part in ensuring that the municipality retained control over materials, doubling up as detectives investigating the illegal diversion of discards away from Felipe Cardoso, for clandestine disposal, unregulated recycling, or illicit consumption. I was present when they received a complaint alleging that a garage in upper-class Carrasco had been dumping tyres at the back of the Felipe Cardoso shantytown, presumably paying residents a small fee but avoiding the larger cost of shredding the tyres and depositing them in Felipe Cardoso. Invisible from the street and difficult to access for inspectors, a large pit of tyres was nevertheless visible in a Google Earth print-out that the Laboratorio staff studied in the office. Another day, Joana mentioned the name of someone who had been picking up waste from a large furniture company without the appropriate documentation, and without then taking it to Felipe Cardoso. Was he perhaps a clasificador recycling materials in the informal sector? The fact that I knew most waste-pickers only by their nicknames meant that I could honestly answer that I had no idea who Joana was taking about.[11]

Waste-picking activity was, for municipal bureaucrats, shadow-like. Often unmentioned in official paperwork, waste-pickers were present in places where they should have been absent, operating at the margins of legality and regulation. Nevertheless, they had to be accounted for both in everyday bureaucratic activity and in longer term infrastructural planning. The image of clasificadores labouring in the shadow of Felipe Cardoso's waste mountain provides the perfect illustration of the adaptation of municipal infrastructure to waste-pickers' presence. But if there was municipal oversight at the landfill, the informal labour of clasificadores was elsewhere difficult to grasp, bureaucratically illegible, ghost-like even. This was a world of blurred photographs of wrong-doing, phantom 'companies', which operated without municipal registration, and unconfirmed reports of illegality.

As Josh Reno (2016) sets out in his ethnography of a Michigan landfill, the 'enabling activity' of waste infrastructures lies in the way that they allow people to live their lives in relative separation from the discards they co-produce. Clasificadores, in their extraction of waste from municipal bins, in their organised collections from neighbours, and by extending the life of the landfill through reducing the waste buried there, exercise this infrastructural function. In the case of waste collection, clasificador infrastructure runs parallel to municipal collection, a situation that is intolerable for Uruguayan urbanists like Baracchini and Altezor (2010: 271), who argue that municipal control over primary services like waste collection is 'inalienable' and should immediately be reasserted over the 'primitive use of little carts'. Yet the infrastructural logic that motivates clasificador activity is clearly different from that which organises municipal collection. Clasificadores focus on recovering value in discards, rather than enclosing and eliminating them in order to minimise the risk they pose.

Beyond Uruguay, we find other groups who seek value in discarded things and operate shadow infrastructures. In Britain, for example, gypsy travellers operated the scrap trade long before formal sector private companies began to take an interest. Of her work with travellers in the 1980s, Sharon Gmelch (1986: 312) writes that 'by scavenging and being alert to the existence of usable objects in their environment' such groups can ensure their own economic futures. The link between travellers and discarded materials continues today: in his recent fieldwork with Irish travellers in London, Anthony Howarth found that men collected materials such as scrap metal and construction wood, and stored such 'resources with a potential re-use value' in their yards (2018: 59). Long victims of legal enclosures that have restricted their freedom of movement, UK-based travellers have recently been further disadvantaged by the Scrap Metal Dealers Act (2013), which brought in the stipulation that anyone trading in scrap would have to register in each of the localities in which they worked. As one scrap dealer put it:

> this infringes the human rights of Gypsies and other Travellers who have traditionally carried on [sic] the scrap metal trade. From metal working in the Middle Ages they have come through hundreds of years – recycling metal, selling metal, using metal, sorting metal – and this law is going to effectively kill off their lifestyles because they will have to pay for a license in every borough or county which they travel though. (Le Bas 2014)

Let us turn to a radically different example where the expansion of state and formal private sector provision has displaced a shadow infrastructure performing a disparaged service. Often, it is assumed that a lack of trained gynaecologists involved in delivery meant that women in the developing world went without support in pregnancy or childbirth. Medical doctors enter as white knights operating on a *terra nullius* of non-existent ante-natal care. Yet such narratives ignore the presence of *parteras* and indigenous midwives whose knowledge and work differs from that of medical professionals, some of whom are clearly beneficial (supporting and empowering women), others less so (providing less specialised support in emergency situations). Such practices are often marginalised through processes of medicalisation that, in the Global South, regularly involve the growth of aggressively marketed for-profit medical services. For example, Murray de López (2015) writes that the provision of free and low-cost health care has expanded in Chiapas (Mexico), but this has contributed to what she calls the 'disappearing midwife', where traditional and so-called 'empirical midwives' are being sidelined by medical professionals. Murray de López argues that the medicalised model of birth in Chiapas 'openly rejects midwifery as authoritative knowledge on the subject', leading to a situation where '*parteras empíricas* are disappearing both in social presence and from reproductive health discourse' (2015: 18).

A final example of a shadow infrastructure at risk from formalisation is that of the informal lender. Like the expansion of gynaecological health care, the expansion of credit and lending to those at the 'bottom of the pyramid' is mostly uncritically celebrated. Yet my research with waste-pickers in Uruguay adds nuance to such celebrations. Waste-pickers working in the informal sector were regularly granted interest-free loans by the intermediaries who bought their recyclables and whom they often paid back in kind. When their labour was formalised, however, they were able to access formal credit markets and often found themselves heavily indebted at high-interest rates to pay-day lenders. This situation made me wonder whether similarly unexpected developments with debt might not be occurring elsewhere. This seems not to be the case in India, where informal lending is an option of last resort for the poor (Kar 2018). Yet anthropological research such as Deborah James's (2014) work on financial inclusion in South Africa suggests that a rethinking of the stereotype of the disreputable loan shark might well be in order. In the South African case, so-called 'credit apartheid' meant that informal lenders – known as *mashonisas* – often lent money to black South Africans at more favourable rates than whatever commercial loans might be available (2014: 111). Further, such lenders often emerge following pressure from borrowers, and are much more embedded in their communities, which place social constraints on interest rates (James, 2014). There is no consensus, therefore, that the opening up of formal sector credit to black communities is necessarily beneficial to borrowers, but it does move profits from community-based black *mashonisas* to white-led financial companies.

What do these disparate cases of gypsy scrap dealers, indigenous midwives, and informal money-lenders briefly sketched out here share in common? They represent instances of socioeconomic niches that corporations had considered of little interest, leaving a shadow infrastructure to provide a service to local communities. Then, changes in political direction, the legislative environment and, ultimately, bottom lines, encouraged large companies to move in and partially displace such an infrastructure. In the cases of the poor Mexican women and black South Africans, such groups may have been treated as barren wastes by state and corporate elites, and the state-led democratisation of access to maternal health care and credit are generally regarded as progressive moves. Yet they also have unintended consequences, opening up spaces for capital accumulation and displacing latent forms of knowledge and economic activity. In all three cases, operators of the shadow infrastructure are denigrated in order to justify their expropriation: *parteras* are dismissed as dangerous 'witches', the Scrap Metal Dealers Act is said to tackle 'rogues', and the *mashonishas* are portrayed as violent loan sharks. As we shall see in later chapters, the same 'epistemological violence' has been applied to Uruguayan waste-pickers working in the informal sector.

Conclusion

This chapter has approached Montevidean waste primarily from a municipal perspective. While this is in part a pragmatic choice (these are the records that are available), more importantly it is because municipal authorities, in common with many parts of the world, have long held responsibility for the management of Uruguay's largest waste-stream. Indeed, the emergence of municipal solid waste (MSW) is contemporaneous with the birth of municipal politics in Uruguay's capital. For the most part, municipal waste management in Montevideo has been based on a linear model of classification, containment, transport, and elimination by landfill or incineration. This approach amounts to what I call 'hygienic enclosure': attempts to minimise waste's potential for hygienic and aesthetic risk, and maintain a municipal monopoly on waste management at all stages of the waste disposal process. Late twentieth-century neoliberalisation opened up limited spaces for private sector involvement in containment, transport and indeed the creation of waste, but this was still guided by the granting of municipal concessions.

Together, the minimisation of hygienic and aesthetic risk, and the attempt to establish a municipal monopoly on waste collection can be seen as constants in what I have discussed as the shifting infrastructural modernity of Montevidean waste collection. Within this model, the substance of risk may itself change or remain the same. Germs replaced miasmas as the source of epidemic and hygienic risk at the beginning of the twentieth century, and the foulest smelling industries ceased to be automatically conceived as the most dangerous. With respect to aesthetic risk, on the other hand, there appears to be more consistency, at least during the twentieth century, when omnipresent overflowing bins were negatively perceived as indicators of infrastructural failure and threats to Montevideo's modernity. Another relative constant has been the way in which infrastructural models and visions of modernity have been sought from beyond Uruguay's borders, principally in Europe. Montevideo was one of the first South American cities to install English-built incinerators at the beginning of the twentieth century and, at the beginning of the twenty-first, important political figures lobbied, unsuccessfully for the time being, for Uruguay to become the first in the continent to pioneer a waste-to-energy plant, modelled on the Italian city of Brescia (*El País* 2014). The notion that Uruguay imports ideas and technologies is hardly a revelation, but it has meant that a native shadow infrastructure operated by waste-pickers has often been framed as backward and a challenge to infrastructural modernity. This despite the fact that customary access to waste also exists and faces similar attempts at enclosure in the Global North, from Irish Traveller scrap collectors in the UK, to hobo and homeless recyclers in San Francisco (Gowan 2010).

The fact that municipal waste managers have had to adapt their infrastructure to account for the presence of clasificadores can most clearly be seen in the stipulation that materials entering the landfill must be 'rendered unusable'. This clause, easily framed as an environmental imperative, in fact addresses another kind of risk. Waste-as-discard is transformed into waste-as-unusable-material in order to prevent landfill waste-pickers accessing the waste commons to share, consume, or possibly recommodify 'ex-commodities' (Barnard 2016) or 'would-be commodities' (Boarder-Giles 2014: 12) on their own terms. Clasificadores and municipal waste-workers share a common infrastructural function when they collect discards from citizens – allowing them to live separated from their waste. Yet the social relations and spaces for capital accumulation enabled by municipal decree are centred on practices and technologies of containment and destruction, while those of waste-pickers liberate the potential value and relations embedded in discarded things, as the following chapters explore.

2
The Mother Dump
Montevideo's Landfill Commons

A desalambrar, a desalambrar!
que la tierra es nuestra,
es tuya y de aquel,
de Pedro y María, de Juan y José.

Let's tear down the fences, let's tear down the fences!
this land is mine,
yours, everyone's,
Pedro and María's, Juan and José's.

'Desalambrar', Daniel Viglietti, Montevideo, 1968

I was a few months into my longest period of research with Uruguay's waste-pickers and life at the COVIFU housing cooperative, a few blocks away from the Felipe Cardoso landfill, when my next-door-neighbour Juan gave in to my pleading and allowed me to join him for work at what waste-pickers referred to as 'the quarry' (*la cantera*). I woke with the sun and joined him and his wife María for a few morning *mates*. Juan and María's five young children were still asleep, and the couple enjoyed a quiet moment of peace and intimacy before he left for work and the children began to stir. Following the *mates*, Juan and I each hopped on to our scooters, kickstarting them down the red earthen path that led us past a few whitewashed homes of the housing cooperative, patches of trees, tied-up horses, and an abandoned brick kiln. Juan's mother and stepfather's house was the last we passed before the path turned into the asphalted but potholed road leading to the landfill. We overtook a few trucks that were making their way to the dump but it was early, and traffic was yet to build up. In the blink of an eye, we were back on a dirt track, this one better maintained by the Intendencia, an internal road linking different parts of the complex of open and closed landfills, and their associated gas and leachate treatment plants, that form the waste complex of Felipe Cardoso.

On the opposite end of the landfill from the weighing station where the trucks enter, we hid our scooters in a patch of trees, where they joined other motorbikes, a few cars, and a horse and cart. We entered through a gaping hole in the landfill's fenced perimeter. The path inside was initially flat and dry, then became wetter and steeper as I followed the confident steps made by my friend as he made his way across an array of bright plastics, truncated tree trunks, thick tangled fishing ropes, and a thousand other discards. When we reached the top of the mountain, we were able to make out figures sifting through the garbage and hauling large sacks onto their shoulders. We were far from the reversing dump trucks, municipal workers (*municipales*), police (*milicos*), and other waste-pickers. The price of PET at that time was generous enough for Juan to make do with scouring the old waste for plastic bottles, and I helped him with this task. Alongside these, he also kept a sack for metals and *requeche*. Into this bag he threw items of clothing and jotters for his children, a pair of shoes he thought I might like to give to my girlfriend, and a few unfinished bottles of soft drink.

Juan and I share ages and a passion for football and, on that day, we were both dressed in similar old clothes that we were willing to get dirty: dump clothes. Black sports caps protected our heads and faces from the sun. Yet enough of my face was on show to reveal that it was a new face, and soon enough Juan's uncle shouted over, seemingly to complain about my presence. Juan instructed me on the response I should volley back: 'the landfill has no owner [*la cantera no tiene dueño*]!'

Following a consideration of the way that my clasificador neighbours, workmates, and interlocutors interact with and describe the landfill, this chapter suggests that rather than a dystopian nightmare, the space can be understood as part of an urban waste commons. For clasificadores, the *cantera* was 'a giant playground', the 'big free shop', and a 'mother' to whom they could always turn. After situating my work in the context of wider commons scholarship, I focus on the activity of so-called *gateadores* at the landfill, detailing their resistance to attempted 'hygienic enclosure'. I then transition into a comparison between the historic English commons and what I postulate can be understood as the contemporary urban commons of Montevideo's landfill before zooming out from the Montevidean case to suggest how the argument is relevant beyond the Uruguayan field site.

The chapter also seeks to conjure something of the spirit of Mayhew, with regard to what Eileen Yeo notes as his foremost insight:

> that people are not to be treated as averages according to whether they fall above or below some quantified poverty line. The communal traditions and memories of groups of people, their aspirations and fears have to be understood before policy-makers draw up their plans for others. (1971: 95)

I choose to define Montevidean waste as a commons based on comparison with the classic case of the rural English commons. While hugely varied and particular in terms of property regime, landscape, usage pattern, resource and management, drawing on the work of historians such as J.M. Neeson, E.P. Thompson, and Christopher Rodgers, I identify several characteristics which hold for these commons in most cases. The first is the fact that commons are not equivalent to open-access resources (many believe that Hardin confused the two) and are often claimed by particular vulnerable populations and subjects, such as widows, children, and landless peasants. The second is that there is a strong emphasis on the extraction of materials for use-value or subsistence, without this precluding gathering or collection for market exchange. The third is that both rural commons and the waste-stream provide refuges from wage labour. The fourth, connected point, is the blurring of the line between work and play in both commons. Finally, and importantly, in both cases we encounter processes of enclosure and resistance to them.

There is a danger in appealing to a concept that has become increasingly expansive, accommodating 'artificial commons', urban spaces, and seemingly every public or semi-public resource at risk of privatisation. As Silvia Federici (2010: 4) notes, much radical commons theory also focuses on 'the formal preconditions for the existence of commons' rather than the 'material requirements for the construction of a commons-based economy enabling us to resist dependence on wage labor and subordination to capitalist relations'. Lauren Berlant, (2016: 395), meanwhile, cautions that, 'although the commons claim sounds positive, it threatens to cover over the very complexity of social jockeying and interdependence it responds to'. I share these concerns over the concept's expansiveness yet also fear an ideological narrowing of what might constitute a commons, something I think is inherent in David Harvey's (2012: 73) definition of commons as 'both collective and noncommodified – off limits to the logic of market exchange and market valuations'. Adopting his position would clearly exclude Montevideo's landfill as a commons, since the waste-pickers working there make most of their income from the sale of plastics, paper, cardboard, and metals in a capitalist, often transnational marketplace.

Montevideo's waste commons, I suggest, has more 'in common' with the classic case of the English commons than with the square occupations or online activism (Levine 2002) that have dominated recent commons scholarship (Harvey 2011; Susser and Tonnelat 2013). I opt to return to the roots that sustain later commons scholarship, combining a social history of the heterogeneous English commons with oral history and contemporary ethnography of Montevidean waste-pickers. One advantage of focusing on the English commons is that they are the first and one of the few cases where territories were explicitly called commons. That is to say, other patterns of land use and ownership, in India or Mongolia, say, are described as commons either by comparison to the English case, or, more habitually, by reference to a definition of the commons ownership, such as that developed by Elinor Ostrum and her colleagues (Ostrum 1990). In this process, commons inevitably become stripped down, losing some of the heterogeneity that make the diverse English commons so good to think with. Against this and the ideological purifying of scholars who focus on the absence of exchange and commodification as the sine qua non of any commons, I suggest that the economics and management of urban commons like Montevideo's landfill are inevitably 'messy' without this precluding the existence of important continuities with traditional rural commons.

Felipe Cardoso: the (not so) final resting place

As we have seen, Montevidean waste disposal oscillated between incineration and landfill during the course of the twentieth century. Incineration was initially replaced with a series of short- and medium-term dumps that filled in natural depressions and quarries in different parts of the city until, in the 1970s, the Intendencia began to concentrate waste disposal in the east of Montevideo. When I interviewed a key municipal waste engineer, he extolled the advantages of the area for landfilling, including low population density, clayish soil, and a considerable distance from the city's water supply. Some maps still refer to the area as Las Canteras (the quarries) due to the large number in the area and, according to the engineer, dumping began when local brick manufacturer Andrés Deus agreed to let the municipality fill some of his quarries with waste.

A little detective work brought to light a more serendipitous account of the founding of the *madre cantera* (mother dump) at Felipe Cardoso. When the Intendencia returned to landfilling in the 1960s, this was by no means meant to be a permanent solution. In his 1958 report, Francisco Bonino advocated the 'fertiliser' or 'incinerator' disposal method instead, arguing that landfills, which 'always create environmental problems because of imperfections in compaction or cover', should only be reverted to 'in emergencies and on the smallest scale possible' (1958: 16). It was thought that the filling of geographic depressions and quarries would soon reach a natural limit, and the activism of neighbours in surrounding areas further restricted municipal room for manoeuvre as occurred in New York (Gandy 2003) and Buenos Aires (Suárez 2016). As we have seen, neighbours objected not just to odours, flies, rats, and consequent risks of disease, but also to the presence of waste-pickers. A representative of a neighbourhood association near the Lussisch quarry told one newspaper that the neighbours didn't want to 'continue fighting against the Cantera de los Presos shantytown, whose residents say that they will set up home wherever the rubbish is taken' (*Hechos* 1965).

The Intendencia was also unhappy with the presence of waste-pickers at its dumps. At a 1968 press conference, which celebrated the (much contested) success of Operativo Limpieza, the municipal head of Engineering and Public Works said that the Intendencia was trying to 'prevent groups gathering at dumps ... in order to collect newspapers and objects that they subsequently sell' (*Acción* 1968). The situation was 'particularly alarming at the so-called Cantera de los Presos, where people expose themselves to serious risks because of the sanitary conditions' (*Acción* 1968). The problem, he claimed, would be 'definitively solved with the creation of Usina 5, which would soon be able to transform, daily, around 400 kilos of rubbish into fertiliser' (*Acción* 1968). The plant at Felipe Cardoso was to be the first of the Intendencia's *usinas* to be dedicated not to incineration nor to landfill but to the transformation of waste into compost.

The initiative originated with a group of councillors belonging to the Frente Amplio (*El Popular* 1972), whose first incarnation enjoyed a brief existence before being outlawed by the dictatorship. Optimistic accounts suggested that the sale of the fertiliser could earn the Intendencia large sums of money (*El Diario* 1972) in the context of leftist proposals to institute import-substitution industrialisation that would replace fertilisers bought from overseas. But the plant, whose machinery was itself imported from England, was plagued with difficulties from the outset.

'The machinery was installed', reported one visiting journalist in 1971, 'but [the plant] could not and cannot begin to function because the plans for the main switchboard were missing' (*La Mañana* 1971). The composting plant was still the municipality's 'great hope for the daily elimination of a large volume of waste without endangering the health of the population' (*La Mañana* 1971). But the 'acrid and penetrating smell' that 'invaded the lungs' of the journalist in question came not from fertiliser but from the 'University Campus Cantera' a few blocks away, where he reported that 'the city's waste [was] currently dumped' (*La Mañana* 1971). When the plans finally arrived from England, a power-cut further delayed the *usina*'s inauguration (*Ahora* 1972), and when production began, doubts persisted about the quality of a product that, due to its deficiencies, could not in fact be marketed as fertiliser or compost, only as 'soil improver' (*El Día* 1973a, 1973b).

Production lasted barely a year, and the site very quickly became yet another landfill, exemplifying AbdouMaliq Simone's point that 'the apparent systematicity of cities is in large part a process of "one thing leading to another"' (Venkatesan et al. 2016: 16). 'And the rubbish keeps piling up' headlined an instructive 1973 article, which noted that while the *usina* 'wasn't working', the installation and its surroundings were 'completely filling up with rubbish' (*Ultima Hora* 1973). Neighbours complained that the place had become 'totally unhygienic and unwelcoming because of the immense quantity of unknown individuals who marauded day and night in the neighbourhood, attracted by the enormous mountain of rubbish' (*Ultima Hora* 1973). These 'marauders', none other than the waste-pickers who form the subject of this book, had even taken over the sanitary facilities built for municipal workers (*Ultima Hora* 1973), which they continue to use to this day. Usina 5 thus only 'solved' the waste-picker problem in that it gave them a semi-permanent space where they could labour intermittently over the following decades and up to the present day.

The official title of Felipe Cardoso today is the Sitio de Disposición Final: the 'site of final disposal'. While it is sometimes referred to as a sanitary landfill, it does not meet the international requirements to be recognised as such (Thurgood 1999), because it is located in the city, is not covered with earth on a daily basis, and features the presence of waste-pickers. As the landfill foreman himself put it, 'for anyone who knows sanitary landfill, this doesn't really compare'. More broadly, Felipe Cardoso can be considered a 'waste complex' that includes the now inactive Usinas 6 and 7; the old Usina 5 area; a leachate and biogas treatment plant; a semi-public area for small-scale dumping in skips; a new private landfill for industrial waste managed by the Uruguayan Chamber of Commerce; nearby private waste transfer and treatment stations; semi-clandestine dumping sites; and the informal housing settlement (*asentamiento*) where many of the waste-pickers live (see Figure 5). All these sites are connected by flows of trucks, waste materials, and people in an example of what Zsuzsa Gille (2015) calls 'waste-dependent development', through which the shallow excavations of brick magnate Andrés Deus and his workers were replaced with the landfill of Felipe Cardoso and its waste-pickers (Figure 5).

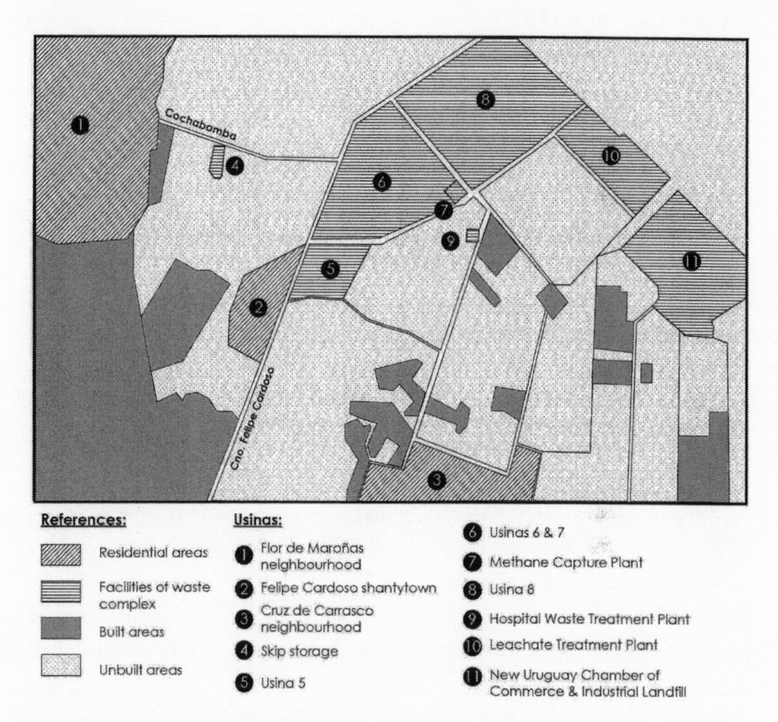

Figure 5 The Felipe Cardoso waste complex

Source: Map designed by Mary Freedman and Adriana Massidda.

Over 2,000 tonnes of waste are dumped at Felipe Cardoso on a daily basis by a vast array of different vehicles. These include the 30 municipal and 7 CAP trucks that come and go with around 1,000 tonnes daily of household waste, leaves, and pruning, but also many other contractors, large and small. Montevideo's largest skip companies – Atlas, Cavok, Volquemax – dump most of their construction waste here, as do faith-based NGOs such as the Organización San Vicente and Tacurú that collect rubbish and sweep streets in Montevideo's shantytowns and popular neighbourhoods. Then there is the multitude of companies that collect commercial waste, from well-established large companies like Bimsa to much smaller operators that have moved into the market following the Intendencia's formalisation of commercial waste collection. The Intendencia currently lists over 170 accredited 'waste transporters', some of which are large companies that generate and then dispose of their own waste (IM 2020). Such trucks are all charged by the Intendencia for each ton disposed of, at a current rate of just over US$40 per ton.[1]

Clasificadores kept a close eye on these vehicles, familiarising themselves with frequencies, timetables and loads so as to ascertain what might be of value. Many of the core group of landfill waste-pickers, known as 'crawlers' (*gateadores*) for the way they had to sneak in under the fence in the past, were kin, and around half lived in the informal housing settlement established soon after the landfill was established in the 1970s. The *cantera* had several respected authority figures: El Ruso, Juan's paternal uncle, a once feared jailbird who worked quietly with his daughter and daughter-in-law (the only two women at the dump at the time) until recently passing away; Juan's maternal uncle, 'El Puto', who had worked there longest; Negrito, one of the few who had persisted at the *cantera* during the 2000s as the Intendencia attempted to move all the *gateadores* to an approved cooperative site; and Enrique, younger, but brought up at the landfill, a man who barely missed a day's work and had amassed significant capital through hard work and access to valuable trucks. Many of the *gateadores* were, like Juan, heads of large families. Another group, loosely referred to as *los gurises* (the kids), were young men like Juan's brother-in-law Leo, who worked less arduously to earn a smaller amount of money that they mostly spent on clothes, trainers, and going out to the dancing.

Gateadores often worked in pairs as partners (*socios*) who could help each other hoist large bags of recyclables (*bolsones*) onto shoulders, keep each other company, and split earnings (a useful safety net if one became ill). Working practices varied between scouring old waste and accessing the fresh waste recently dumped by trucks before heavy machinery (*las maquinas*) arrived to compact and spread it over the dump. The most valuable of these trucks carried not household waste but that from factories, building sites, and large commercial enterprises. *Gateadores* generally tried to complement each other by focusing on different materials (principally cardboard, plastic, paper, and metals) if possible, and switched depending on market value – Juan changed from PET bottles to cardboard during the year – supplementing income with the recovery of metals like copper, steel, and aluminium (Figure 6). For most *gateadores*, the differentiation of materials was a sensory and principally visual activity – a process of training one's eye to recognise value in the trash. While fights over access to particular material streams and trucks had apparently been common in the past, I never witnessed any during my ethnography – in no small part because any violence could have risked everyone's access to the dump.

Recovery implied both a movement from the general to the particular (from mass waste to singled out thing) and from the particular to the general (from a thing with a particular history and characteristics to a general category). Thus, a bronze tap spotted at the landfill would both be differentiated from surrounding *rejectia* and assimilated into a general category (bronze), where importance would be placed on certain characteristics (weight, purity) but not its status as a tap or its life history. Through this training of the eye, a homogeneous pile of rubbish was reconceptualised as a stack of valuable material categories interspersed with trash. At the same time, categories that might operate in other spheres of life were disregarded. Receipts, printer paper, a jotter, a new book, bus tickets all became 'white' (*blanco*).

Part of the sensory engagement involved being attuned to the dangers and risks of the landfill environment. There were two principal risks: an accident involving machinery, or food poisoning. I did not hear of any serious cases of the latter while working at the dump and *gateadores* assured me that through identifying the source of foodstuffs (and therefore which companies regularly contaminated their waste to avoid human consumption) and visual and olfactory tests, they could minimise risks. In terms of the former, only one case occurred. A young *gateador* who was addicted to *pasta base*[2] dived down to recover a bag of metal from the jaws of a compactor, lightly injuring himself in the process. The older clasificadores, who had families to support, were apoplectic with rage at the risk to their common access that such carelessness occasioned, and they banded together to exclude the young man, at least until he had overcome his addiction. In general, *gateadores* defended the safety of their own activity compared with that of waste-pickers in recycling plants, as in the following account from Negrito:

> In the plants and cooperatives, they break open little bags [*rompen paquetes*] but we don't. We have contact with the rubbish because we walk on top of it. But we all have suitable clothing: boots, gloves. Nobody wears normal trainers or sandals on top of the rubbish. We don't break open the rubbish [*romper basura*] or chemical things, or things from the hospitals. We work with things that are useful for us: it's more hygienic. The problem is that people don't know how we work. In the plant, they have a conveyor belt, where people open bags which could have something contaminated inside. That's the issue. (see O'Hare 2018)

Once classified, waste-pickers would sell the materials on to buyers (*compradores*) who came to collect them. Some clasificadores doubled up as small-time *compradores*, buying materials from their colleagues at the landfill; others built up long-term relations with particular buyers – a practice which they felt led to more preferential rates – while others still shopped around for the best prices. Paper and cardboard mostly stayed in the national economy, working its way up from the informal to the formal economy to be transformed by large manufacturers like IPUSA, while plastics often travelled contraband to be processed abroad.

Figure 6 *Gateadores* removing metal from the valuable LAISA steel forge waste truck

Source: Author photo, 5 April 2014.

Commoning

Vulnerable claim-making

Not all of the materials recovered from the waste-stream by clasificadores are sold on to industry. Waste in Montevideo also allows for foraging food, fuel, building materials, and other objects of use and exchange value. Further, not all citizens were able to claim equal access to the bounties of Felipe Cardoso. Let us recall the title of this book – 'rubbish belongs to the poor' – that had been shouted by waste-picker families at the entrance to El Cerro. I want to highlight the nature of this claim, which is not that waste belongs to everyone or should be open access, but that it belongs to a particular constituency: the poor. This is one of the statements which inspired me to think of waste as a commons. For although the normative legal framework for commons in England consisted of an agreement between landowning lords and tenants, by the seventeenth century, it was marginal members of society who were mostly strongly associated with land on the margins (Rodgers 2016). If one looks to the rules and customs surrounding access to and extraction from commons, we find that these link to specific vulnerable groups.

Being grounded in custom rather than law, many use-right claims went unwritten, and could only be glimpsed through moments of conflict, when customary access was denied. Yet some written rules do exist. Peter Linebaugh (2008), for example, has rediscovered a forgotten clause in the English Charter of the Forests (1225) that granted widows the right to the 'estovers' of the common: wood that could be used for house repair and fuel. Gleaning, another form of commoning, can be traced back, among other places, to the Bible, with Deuteronomy (24:19) containing the injunction that 'when you reap the harvest in your field and forget a swathe, do not go back to pick it up; it shall be left for the alien, the orphan and the widow'. The right to glean was defended in the English Commons Pleas court, ultimately unsuccessfully, in 1788, but gleaning was thought to have continued long after (Thompson 1991: 139). In France, a royal edict promulgated in 1554 – and still in place today – established gleaning rights for 'old people, amputees, small children and other persons who lacked the strength or faculty to work' (Sargent 1958: 100).

Like these examples, the *cantera* of Felipe Cardoso has historically provided sanctuary for vulnerable subjects, just as the waste of Victorian London's streets provided a livelihood for rural Irish migrants pushed out by famine or enclosure (Mayhew 1968 [1851]: 139). First, there were migrants who came to Montevideo from rural areas outside of the capital and who either hadn't found other work or went directly to the dump. An older informant, Canario Ramón, told me that he had built the first shack in the Felipe Cardoso shantytown in the early 1980s, having migrated with his mother and siblings from rural San José. The settlement began to grow steadily, initially populated with his relatives. Women and single mothers like Juan's mum, Gorda Bea, could also work at the dump while keeping an eye on their children. Selva first appeared at the Isla Gaspar dump in 1965, when her husband was away in the military and she was struggling to put food on the table for her children (Figure 7). 'I started going to the *cantera*', she told me, 'because we had nothing.' A friend had told her about 'a place where it isn't so bad, where people work and "gather" [*juntar*]'. Intrigued, she turned to her friend, Juan's grandmother China Tore – 'China, let's go to Isla Gaspar, where I've heard that they are dumping and you can make money.' Together, the two women would walk 7 kilometres to the Isla Gaspar *cantera* and back with a four-wheeled cart in which Selva carried her son, niece, and nephew. They would find food to eat and a slaughterhouse truck driver would give them fresh meat. In order to avoid the daily trip home, they began to sleep overnight at the *cantera* with their children, first in a tent, then a shack of sheet metal. There were

Figure 7 Selva in front of her home on Camino Felipe Cardoso, alongside her grandson and holding a portrait of her grandmother, who emigrated from Catalonia

Source: Author photo, 19 October 2014.

no police stationed at Isla Gaspar, Selva told me, and clasificadores could work without harassment. Waste-pickers called this the *cantera libre*: the free dump or quarry.

Those wanted by the law, either for criminal or political activities, also sought refuge: my octogenarian neighbour Carceja, for example, who was persecuted during the 1970s by Uruguay's military junta for being a Communist. 'They wouldn't let me live,' he told me, and the dump was the only place that could offer him refuge, food, and enough money to provide for his children. In the past, older clasificadores told me, wanted political activists from the armed revolutionary Tupamaro movement slept in the woods nearby and would sneak into the dump in the early hours to gather together just enough food for their daily stew. Finally, marginalised and discriminated against Afro-Uruguayans also made up a significant proportion of *cantera* clasificadores. These were figures like Antonio, the father and stepfather of the Azucarero siblings, my neighbours who worked afternoons at the landfill. Now in his fifties, Antonio told me – amidst the beating drums of a family party – that he had finished secondary school in the 1970s but couldn't stand the everyday racism he encountered as a ticket inspector on the buses. So he turned to the *cantera* instead. 'The dirt from the *cantera* makes us all black,' Ruso once told me, asserting a non-racialised space of exception. Yet, as I have argued, the *cantera* was not open access, and the conditions of eligibility for entry were continually debated among clasificadores and involved such factors as need, geographical proximity, and kinship.

There is an argument to be made, then, that the *cantera* functioned as a safety net for those traditionally excluded from social security. Although George Pendle (1952) points to Uruguay as Latin America's 'first welfare state', with a raft of social protections and insurance brought in for workers and citizens by the Battle government of 1910–20, this welfare system, as well as rising and falling with economic cycles, led to a new type of social stratification, consolidating middle sectors, such as public servants and protecting some subordinate groups, especially industrial workers (Filguera 1995: 2). Such social security did not adequately cover women, racial minorities, rural migrants, ex-convicts, or the long-term unemployed.

These vulnerable groups conceptualised the landfill as a giving mother: the *madre cantera*. In referring to the landfill as a *cantera*, clasificadores not only referenced the fact that Montevidean dumps had often been situated at old quarries but indicated their extractive relationship to the space. Indeed, we might say that Felipe Cardoso is land to be filled for the municipal authorities but a quarry to be mined for the clasificadores. Key to the parent metaphor, meanwhile, was the idea that whatever you had done, you could ostensibly always rely both on your mother and on the *cantera* to provide you with at least a plate of food. The first thing Ruso's brother Sordo did on leaving prison was make for his maternal home, and then for the *madre cantera*. Ruso spoke of *requecheros*, who would turn up at the dump with a pot, fill it with food, and leave and he referred to Felipe Cardoso as the 'mother of the rubbish'. 'She was everyone's mother', another informant and COVIFU neighbour Pelado explained, 'because you went there and rescued something to eat, somewhere to sleep, with sheets, mattresses, and no-one would bother you.' In and out of care and foster homes, Pelado had eventually found his way to the *madre cantera*.

For the Uruguayan case, I present the link between custom and waste-work in largely positive terms: the right to access and work waste a privilege to be defended. Yet it is important to recognise that there are other times and places where customary waste-work is seen by the underprivileged as an unjust burden imposed upon them, the most obvious being polluting caste work in India. To take just one example, Gooptu Nandini (2001) details how, in the 1920s, a Dalit council of elders in Kanpur decided to overcome an established custom whereby the untouchables, owing to their low status, were considered to be entitled only to left-over and rejected food and goods of higher castes, called *utran* (discarded, used, or second-hand clothes), *jhutan* (left-over or half-eaten food) and *phatkan* (refused or extra bits). Under the initiative, sweepers refrained from eating *jhutan*, which they used to collect and consume during the cleaning and sweeping. Although the relationship between stigma and waste is important, and will be discussed in later chapters, it can also be problematic, leading us to assume that waste-workers necessarily feel stigmatised. In his contemporary ethnography of waste-work in Lahore, Waqas Butt (2019) rejects the lens of stigma, purity, and pollution, arguing that these obscure, first, the historical and social conditions out of which waste-work has emerged and been organised, and, second, how rural immigrants have played an active role in seeking out waste-work as a way of finding an economic and residential niche in landscapes fraught with long-standing forms of social stratification and exclusion. This is a pattern that I find repeated in Montevideo, in the stories of Canario Ramón, Gorda Bea, and others.

Extraction for subsistence

Much of what waste-pickers extracted from the *cantera* was used for domestic subsistence. After a day at the landfill, Juan would arrive back home like Father Christmas, spilling the contents of a large sack out onto the living room floor to be perused by his excited wife and kids. Soft drinks, biscuits, colouring books, chicken, shampoos, beers, fruit, vegetables, and yerba mate would tumble out: food for his family and pigs, but also to share, so that my fridge was often overflowing with miscellaneous bags of foodstuffs passed over the fence. There were different types of firewood, sheet metal to repair his horse's stable, and colourfully printed cloth that his wife washed, cut, and hung as curtains. Many neighbours made their fences – as well as the pens in which they kept their pigs – from pallets, while the grills on my windows were soldered from metal Juan had recovered from the dump. Jessica, Ruso's daughter, told me that her family only had to buy bread and milk, with the rest of the week's necessities taken from 'the big free shop'. In the absence of a forest ecosystem that would provide them with their needs, clasificadores turned to *requeche*, the 'leftovers' of urban life and industry.

Alongside domestic consumption and storage, *cantera* extraction also enabled sharing among those who otherwise had few material resources. First, without expending too much energy, *gateadores* would often put aside material that they were not themselves collecting but knew that others were. For example, when raking through the valuable rejects from Uruguay's only steel forge, a *gateador* collecting non-ferrous metals might throw any ferrous metals that he found into the pile of a colleague collecting scrap. Second, *gateadores* often set aside *requeche* that they thought another worker might like, tossing it to them with the question 'Any use to you?' ('*Te sirve*?'). Examples of this are cosmetic products offered to women; machine parts offered to someone with an interest in repairs, joinery, or motor vehicles; fodder offered to those with animals; or, in my case, books (I received a fine collection of Uruguayan literature from El Ruso). Third, there was the regular, immediate consumption of recovered foodstuffs such as MacDonald's burgers, ice-creams, nuts, drinks and biscuits. Finally, when a clasificador chanced upon a large amount of a particular *requeche* – a crate of beer, a carton of yerba mate tea, a box of sausages – they would often share these out with their *socio*, other *gateadores*, or extended family members. Temporarily decommodified, abandoned by its previous owner, waste could be shared easily, just as it had been by my dumpster-diving comrades in St Andrews. Of course, it might also be recommodified, and this is what happened to bulk categories of recyclables such as paper, plastic, and metals, as well as to much *requeche* that was taken to the large flea market of Piedras Blancas to sell on the weekends.

As I have noted, the fact that commodities emerged from the landfill would seemingly disqualify it as a commons for scholars like David Harvey (2012: 73), who argues that these should be 'both collective and non-commodified – off limits to the logic of market exchange and market valuations'. To a certain extent, Linebaugh (2014) understands the English commons in these terms and Goldstein (2013: 364) points out that the use of resources from the English commons was almost always restricted to use at home and sale in local markets. Nevertheless, from the Canadian high seas to the English fenland, there has very often been a patchwork of use and exchange value extraction from the commons. 'The fuel, food and materials taken from the common waste', writes Neeson (1993: 158) 'helped to make commoners of those without land, common-right cottages, or pasture.' But it also enabled 'a means of exchange with other commoners and so made them part of the network of exchange from which the mutuality grew' (1993: 158). I therefore consider it important that the presence of market exchange should not preclude the classification of spaces like the *cantera* as commons, not least because such a move would require a kind of ideological and practical purity that is unavailable to most of the world's poor.

Refuge from wage labour

Another characteristic that the *cantera* shares with traditional rural commons is that they allow for a refuge from low-paid wage labour and a fall-back option in case of unemployment. 'If you're sacked, you can always go and make money in the *cantera*', another neighbour, Gabi, explained. When I first met Juan in 2010, he worked at a biscuit factory but was subsequently injured and decided to return to the *cantera* instead of accepting a job in another factory further away. Jessica had worked in several jobs in the private sector, but when she was fired from her last after a dispute, she likewise headed to the *cantera*. Ruso himself, on losing his job in a plastics factory, had decided, in words repeated by many other clasificadores in similar circumstances: 'Well then, I'm off to the *cantera*!' (*'Bueno, me voy pa'la cantera!'*).

Thompson concludes that commoners 'in some part of their lives still felt themselves to be self-determined, and in that sense free' (1991: 179). Such sentiments were echoed in a common refrain from landfill waste-pickers who valued the autonomy of boss-less work. Negrito, for example, told me that should the Intendencia offer him a job street-sweeping or working in the rubbish trucks, he 'wouldn't accept it. Because here I am my own foreman. I am my own boss.' At the *cantera*, clasificadores did not need to stick to a regular schedule and work day, obey orders, or work for someone else. These were all important benefits that they often formulated as 'at the dump no-one tells you what to do, you don't have a boss'. The lack of fixed working hours meant that *gateadores* could take unannounced days off. Juan, for example, would sometimes stay at home to work on odd jobs, and once had to take time off to look after his kids when his wife was admitted to hospital. In the following weeks, he was able to make up for the lost earnings by working more intensively and for longer hours. El Ruso indulged in periods of binge drinking, when he would be off work for several days, while his daughter took leave when recovering from an operation. Autonomy and freedom from orders were also associated with masculinity, so that workers like Enrique could earn respect and money without either turning to crime or suffering the perceived indignity of service sector work. The *cantera* is all the more remarkable then for its ability to combine a space for the realisation of masculine identity with the historical provision of labour for women who needed to work but could not afford childcare.

In the Hammonds' (1987 [1911]) classic study, part of the design and consequence of the eighteenth- and nineteenth-century British parliamentary enclosure acts was the forcing of rural freeholders and peasants into the ranks of agricultural waged labour. This narrative has been contested, but E.P. Thompson's (1991) extensive research led him to the conclusion that the commons did play an important role in enabling peasant self-sufficiency up until the nineteenth century, even if such subsistence might not have been 'any more than meagre' (1991: 178). Income at the *cantera*, on the other hand, was anything but meagre. According to my research, landfill waste-pickers in 2014 earned roughly the same in a week from the sale of materials as clasificadores in recycling plants earned in a month. The unskilled jobs available to *gateadores* simply did not match the hours to income ratio possible in the *cantera*, recourse to which gave them greater choice over what jobs to accept. This is in stark difference to the idea of waste-pickers as a sort of 'reserve army of unemployed', under-employed, or lumpenproletariat that could be used to keep wages down.

The possibility of such high incomes was, to a large extent, due to the sheer concentration of waste at Montevideo's landfill and the relatively small number of waste-pickers operating there. Unlike *carreros*, who had to travel around the city and were limited by the carrying capacity of carts pulled by hand or horse, *gateadores* had few limits to the amount of recyclate that they could accumulate and sell on-site, conditions permitting. When allowed access, they had the pick of hundreds of trucks and thousands of tonnes of discards on a daily basis. Certainly, some useful surplus material was intercepted before entering the waste-stream. I witnessed some such instances – baked goods from a catering company distributed as charity in Flor de Maroñas, cakes given to the former clasificadores who collected waste in Casavalle with the Organización San Vicente. Yet far more numerous were the quantities of such materials that made it to the landfill and could be redistributed by *gateadores*, if they could rescue them before they were earthed over.

Playful work and workful play

Further common ground between the historic commons and the *cantera* can be found in the blurring of lines between recreation and work. Goldstein (2013: 265) has noted how children grew up working, foraging, but also playing on the historic wastes, citing Katz's (2004) concept of 'workful play and playful work'. Most of the *gateadores* I spoke to went to the dump as children to engage in a mixture of activities that involved both work and play. For example, one of the main attractions for children and teenagers was the possibility of hunting birds, with Juan a particularly good shot with a catapult. 'At first I went to mess around' ('*Al principio iba a joder*'), countless clasificadores told me when I asked them how they were initiated. The recreational activity of the neighbourhood children was bound up with exploring, hunting, and scavenging, as well as provoking the police and older, drunken clasificadores. *Gateadores* transitioned from play to work at the dump as they moved from childhood, through adolescence, to adulthood. Work had replaced play as the principal mode of action for the active *gateadores* at Usina 8 when I began fieldwork, but residual elements of recreational activity remained. The *gurises*, in particular, would often spend many hours at the dump chatting, joking, eating, and engaging in what they called 'mandarin warfare' – the throwing of soft fruit at each other and older clasificadores. Such an activity had been recorded by Hugo

Alfaro (1971) at Montevideo's short-lived Burgues landfill in 1971, where he encountered young lads 'throwing rotten oranges at each other [and] giving themselves up to the joys of recreation'. Kathleen Millar also identifies a 'blurring of work and play' (2015: 34) at Rio de Janeiro's Gramacho landfill, arguing that this engenders a 'time-sense' in waste-pickers that made transition into waged labour difficult.

Resistance to enclosure

'What is Scotland like?', the *gateadores* often asked me. 'I bet they don't stop you working like here in Uruguay – this must be the only country where they don't let you work!' They were referring to the police presence at the dump and repeated attempts to prohibit their activity, what I call a 'hygienic enclosure' that sparked resistance and feelings of injustice. Before the 1970s, access to other city dumps – known as 'free dumps' (*canteras libres*) – had been unimpeded. Theories explaining the installation of the police at Felipe Cardoso varied – from control of violence, to accessing the best material, to safeguarding the machinery or avoiding accidents. Although some attempts to minimise risk were always present, they became more pronounced in the 1970s and 1980s, with growing awareness of environmental contamination and the public health risks of landfill sites. Enclosure of Felipe Cardoso took place in the context of a move away from open dumping towards sanitary or at least controlled landfill in much of the Global South in the latter half of the twentieth century (see Reno 2008: 88). One several times director of Felipe Cardoso, told me that: 'We tried to turn an open dump into a controlled one, in terms of registering the weight and the number of trucks entering but also controlling the entry of outsiders' (Figure 8).

Many clasificadores could reel off the different police units that they had experienced at Felipe Cardoso, as well as the severity of their reign: the Republican Guard (*La Republicana*), ordinary police (*La Policía*), private security (*La Guardía Blanca*), the police dog unit (*El Plantel de Perros*), and the feared mounted police (*Los Coraceros*). In order to continue working, clasificadores initially had to labour clandestinely, or bribe the police for entry by buying them meat, wine, and cigarettes. Clasificadores from that period described how the police would beat them up, interfere with their work and arbitrarily expel them. Carceja, the persecuted Communist, described how the police would set fire to the piles of materials that *gateadores* had set aside, and to their tents when they assembled them in nearby woods, barely caring if there were children inside. Gorda Bea remembered when eight policemen had been stationed at the dump who 'didn't let you work, chased you, beat you'. At times, police would try to persuade the women to perform sex acts in exchange for access, Selva told me, while at others only women were permitted to enter, leading at least one man to take up cross-dressing. During the 'infrastructure of elimination' of the Uruguayan dictatorship, everyday violence was complemented with police raids on the *cantera* and the shantytowns where the clasificadores lived. Reinforcements would be called, and the men taken to police cells in commandeered rubbish trucks.

The *gateadores* were in a difficult position, since they had to work but could not openly confront the police without being arrested or worse. So, they waged a war of attrition, using 'weapons of the weak' (Scott 1985). When taken to the police station in dump trucks, Carceja told me that clasificadores would fill up their boots with rubbish and empty them into the prison cells, causing such a stench that the police captain would be forced to order their release. To prevent the police from burning their tents, clasificadores would tie them up in trees during the day and let them fall only at night. When guards detained waste-pickers in pits at the dump, Gorda Bea, who boasted of being 'more macho than the men', snuck off to set them free. Knowledge of the terrain was essential for this warfare, especially when dealing with the mounted police. Carceja told me how he and colleagues would lead the pursuing police into parts of the dump where there were trenches or pools of water that horses would refuse to cross.

Suffering repression at work, *hurgadores* were no safer at home. Living in shantytowns (*asentamientos*), they suffered regular police and army raids where men were tortured and detained without charge for periods of days to years. In the countryside, Zuli had barely been conscious of the dictatorship but soon found out about it when she moved to the capital. Living in the *asentamiento* of Isla Gaspar, her clasificador husband was regularly detained and tortured. 'It would be cattle prod and submarine, cattle prod and submarine,' she told me, referring to the techniques of electric shock and water-boarding employed on him. Selva was at the time another Isla Gaspar resident whose husband was detained after being caught with a list of supposedly leftist militants who were in fact members of the local drumming corps. He was tortured and jailed for eight years on trumped up charges and now receives government compensation for a shattered kneecap.

These stories are from the 1970s and 1980s, but much of the same struggle continued into the 1990s and 2000s. In the early 2000s, for example, the *coraceros* were brought back to guard the *cantera* when some clasificadores were allowed to classify on an internal landfill road, Cepeda. Juan told me that at that time he was one of those who most 'made war on the *coraceros*', and his case exemplifies a genealogy of resistance to enclosure. His grandmother China Tore had accompanied Selva when she first went to the *cantera libre* in the 1960s, while his mother Gorda Bea had raised him on the dump and played a prominent role in resisting police violence and breaking enclosure. Struggles with police were then passed on to Leo, a young brother-in-law that Juan and his wife María had raised like a son. Leo described to me how, on one occasion, a group of clasificadores defecated in the police cabin, smearing their faeces all over the walls in a dirty protest at their presence. Then, aged just 15, Leo was shot in the back by a drunken policeman at the dump. The bullet is still lodged near his heart – he pressed my fingers to it in his chest – while the policeman was never brought to justice. Such episodes highlight why clasificador trade union leaders would often complain that while the dictatorship ended for most of the population in 1985, it continued for waste-pickers and those suffering state violence in the shantytowns.

Gateadores, like street waste-pickers, endured varying periods of permissiveness and repression of their activity. Yet in the face of police harassment, many persisted. Negrito told me with confidence that: 'There's nothing else but to accept us because whatever guard they assign, whatever they do ... there will always be *gateadores* in the *cantera*.' In perforating the landfill fence and fighting agents of enclosure, the actions of clasificadores resemble those of English commoners centuries earlier and their famous slogan: 'Down with the fences!' In one memorable case from 1830, cited by Linebaugh (2008), commoners from Otmoor, Oxfordshire 'armed with reap-hooks, hatchets, bill-hooks, and duckets ... marched in order around the seven-mile long boundary of Otmoor, destroying all the fences on the way' (2008: 153). *Gateador* knowledge of the landfill terrain also finds a parallel in the way that, in their struggle for the commons, Thompson (1991: 103) writes, the English 'peasantry and the poor employed stealth, a knowledge of every bush and by-way, and force of numbers'. And when *municipales* and *milicos* skimmed off the best waste to augment their wages, can they not be compared to the 'forest officers and under-keepers, who had long supplemented their petty salaries with perquisites' (1991: 103)?

The enclosure of distinct English commons was generally carried out by private landed interests and given blessing and protection by the parliament and courts in which they often held office. It was justified through a discourse of reinforcing private property, ensuring the nation's food supply, and making wastelands productive and profitable (Neeson 1993: 46). There was an ideological conflation of 'wastes' and wastage in the economic sense as common land was derided as unproductive: as one report to the Board of Agriculture had it, 'common fields may be called the worst of all wastes' (Goldstein 2013: 366). The hygienic enclosure of Montevideo's landfill is clearly different from that of rural land, inserted as it is into a history where waste has been contained as a public bad in order to protect the population from hygienic risk. While peasants were denied access to land because their labour was deemed economically wasteful, clasificadores were to be denied access to the landfill because its contents were classified as unsanitary stuff, 'dead commodities' rather than materials with productive potential. One enclosure sought to transform 'wastes' through capitalist production, another to stop economic activity in its tracks at the landfill gates.

A more familiar comparison can be drawn with the restrictions placed on some of the Victorian scavengers described by Mayhew, many of whom were forced into the city by rural enclosures. Take the case of the 'shore-hunters', for example, who in the 1860s were faced with new restrictions on their access to London's sewers, justified on grounds of hygiene:

> The shore-hunters of the present day greatly complain of the recent restrictions, and inveigh in no measured terms against the constituted authorities. 'They won't let us work the shores', they say, 'cause there is a little danger. They fear as how we'll get suffocated, at least they tell us so, but don't care if we get starved!' (Mayhew 1968 [1851]: 152)

From the perspective of both shore-hunters and clasificadores, who argue that waste 'belongs to the poor' and remember when they accessed it freely, the brutal methods used to exclude them can hardly be explained away by arguments of the 'public interest' kind. The war clasificadores waged to access waste, and the state repression they encountered, mimicked and sometimes overlapped with Uruguay's better-known armed conflict, between Tupamaro urban guerrillas and the fascist military junta. At the landfill, however, the struggle is not for socialism or post-capitalism but simply for the right to work the waste and access its potential for sustaining kinship, subsistence, and even progress. The words of Mayhew, with regard to shore-hunters, ring equally true here: 'it is ... more than suspected that these men find plenty of means to evade the vigilance of … officials and continue to reap a considerable harvest, gathered whence it might otherwise have rotted in obscurity' (1968 [1851]: 152). The question of access to the *cantera* and its materials is in many ways part of the broader question of waste politics and economics: who owns the waste and who has the right to exploit it? I have argued that, like the English commoners of old, clasificadores stake and defend a moral claim based on custom and necessity.

Figure 8 Part of the fence at Felipe Cardoso, with damaged waste containers on the inside

Source: Author photo, 20 October 2014.

Beyond the cantera

What wider relevance does the waste commons thesis have beyond Uruguay? As Medina notes, waste-pickers can be found anywhere where there is 'chronic poverty, high unemployment, industrial demand for recyclables, and … the lack of a safety net for the poor' (2005: 18). Yet not all of these necessarily claim a customary right or are able to access waste as a commons. In certain places, such as in Kenya (Tom Neumark, personal communication 2018)[3] and in some Asian cities (Furedy 1990), landfill waste-picking appears to be controlled by criminal gangs – describing waste as operating as a commons in such circumstances probably would not make much sense. On the other hand, Furedy (1990: 5) notes that 'some waste recovery practices are, in a sense, survivals from a traditional social contract'. The examples she gives of customary waste rights – of Bandung doormen to household waste, Indian leather-caste workers to animal skins and Sri Lankan launderers to coconut shells – bring us much closer to the kind of common-law particulars explored in the work of social historians of the English commons. These are only some of the instances where customary claims to materials rest on their classification as waste; where such wastes can be considered as commons; and where re-categorisation as resource puts access at risk.

We need not only look to the Global South, however, and conditions other than those listed by Medina might also lead to the emergence or growth of scavenging. A fellow activist from Glasgow, for instance, tells me that he used to work recovering materials with his dad in the city in the 1980s. A truck driver who had lost his job as part of Thatcher's deindustrialisation policies, the father-of-four realised that old vans from newly privatised British Telecom were being sold off cheap: he bought one and used to take his son Liam out to find anything of value: old cars that could be stripped down, copper cables, industrial scrap. Explicitly likening the activity of his dad to that of people who used to live off forests before they were fenced off and enclosed, Liam explained that:

> Glasgow in the 1970s was probably one of the most industrialised cities in the world. And yet in the 1980s it wasn't, so where did all the stuff go? Where did all the factories go, the copper, the lead? It was guys like my dad who were driving into the deserted industrial wastelands and picking the stuff up and stripping it and finding what could be re-used.... There was a whole generation of people my dad's age who had worked in these industries, who then stripped these industries down ... and nobody was stopping any of them. My dad was a little bit like one of these birds that cleans a crocodile's teeth. The crocodile was quite happy to let someone clean its teeth, because it would have had to pay to get someone else to do it.

Liam told me that he thought that Glasgow was itself now a much more enclosed city, arguing that it became a lot more difficult to make a living from waste around the same time that private ownership was more strongly asserted over wastelands, empty lots and building sites. Territories and resources can thus go in and out of being a commons, depending on history and socio-political context, so I am not arguing that waste 'is a commons' generally, everywhere. Clearly, the materiality of waste is important to consider when speaking about the circumstances in which it might possibly constitute a commons: it would be hard to see in what circumstances, for example, toxic waste or even certain kinds of non-toxic, non-hazardous but industrial waste might be considered a commons rather than simply a contaminant. Yet often the toxic and the valuable coexist in the same waste-stream, with Julia Corwin (2019) arguing, for example, that e-waste is frequently described as polluting when handled by informal sector actors in the Global South precisely so as to facilitate its capture, now as a resource, by 'responsible', formal sector companies. The process of decommodification implicit in the creation of waste, encourages its appropriation as a commons, while examples from around the world demonstrate that this commons is at risk of hygienic enclosure. 'Increasingly', Chris Hartmann (2018: 565) argues, 'garbage is viewed by municipal governments, development agencies, and private entities as an urban resource to be enclosed.' Hartmann's research takes place in Nicaragua, around the La Chureca landfill, which sustained a community of between 1,500 and 2,000 waste-pickers before it was replaced by an enclosed, sanitary landfill and a recycling plant for around 500 waste-pickers. In her article 'Accumulation by dispossession', Melanie Samson (2015b) has written of the Marie-Louise landfill in Soweto, arguing that informal 'reclaimers' re-cast the dump as a site for the production of value through a combination of intellectual and physical labour, and struggle. In attempting to enclose the dump, she suggests, the local municipality sought to capture the physical materials interred within it, as well as the very framing and establishment of these materials as valuable, while simultaneously erasing the role of reclaimers in these processes. Kathleen Millar (2018), in her work on Brazil's Jardim Gramacho landfill, explains why many waste-pickers viewed the space as a refuge and preferred work there to the low-wage labour intermittently available to them. The landfill was closed and replaced with a sanitary landfill in 2012, depriving some 12,000 waste-pickers of their livelihood, after granting them a one-off golden handshake. In Subic Bay, Indonesia, Elisabeth Schober (2016) writes of how an indigenous group, the Negrito, dispossessed by the expansion of the US naval base into their traditional hunting and gathering territories, were granted in exchange the exclusive right to scavenge the waste generated by the base, a right they were also later dispossessed of. Christian Sorhaug (2012) conducted research with the Venezuelan Warao, who travelled upriver to the El Bote landfill, where they recovered all manner of things to bring back to their community and describes scavenging at the landfill as an extension of Warao gathering practices in the rainforest. Waste ceased to be delivered to the landfill in 2013 and is being redirected to a new, closed, sanitary landfill.

These cases – and we might cite many more – represent contemporary forms of capitalist dispossession and enclosure justified by discourses of hygiene and hegemonic understandings of dignified labour. Clearly the dynamics of access to valuable waste are different in each but in many countries of the Global South I would suggest that we find a dynamic that roughly corresponds to the framework of commoning that I have set out here, and a corresponding attempt to impose hygienic and infrastructural modernity through technologies of enclosure imported from the Global North and designed for cities without waste-pickers. In many cases we find processes of double enclosure, where indigenous groups or peasants, on finding their land and territories enclosed, seek solace and a livelihood at the landfill, only to then find themselves dispossessed of waste as the landfill is also enclosed.

If I have argued that the waste commons thesis primarily applies to what I have called the Global South, what about richer countries? It is clear that there used to be more widespread scavenging and waste-picking in Europe and North America. There is a vast historiography of French chiffoniers who were disenfranchised by the introduction of containers. In the UK, former Lord Provost of Glasgow Bob Winter, born into a large, poor family, recounts that he was a regular 'midden raker' as a boy (personal communication 2015). More Glasgow midden-rakers can be found in Alasdair Gray's (2011 [1981]: 124) magnum opus *Lanark*, where he describes the protagonist observing a raiding party consisting of 'two boys ... bent over the bins and throwing out worn clothes, empty bottles, some pram wheels and a doormat, while a boy of ten or eleven put them in a sack'. A more distant literary connection can be found in Dickens's (1997 [1865]) *Our Mutual Friend*, in which dust-heap scavengers feature prominently (see also Horne 1850).

We need not only look to literature: recall the case of traveller communities and the scrap trade in Britain, where customary rights to access and work surplus scrap are long established. At some point in the post-war period, however, it seems that the combined improved standards of living, municipal waste collection, and sanitary landfilling severed any link between large communities of waste-pickers and landfills in the United States and many parts of Western Europe. Yet elements of landfill-based extraction persisted. In the local 'amenity tip' of the Scottish post-Second World War new town of Cumbernauld, where my mother grew up, for example, several scavengers lived on-site in caravans, helping people to unload their waste, putting things aside for certain people, and even burning waste for the local council. One of these was Danny, a well-known figure in the community who would set aside any toys and clothes he found for needy members of the community, perhaps those who had been made unemployed by Margaret Thatcher's industrial and labour policies.

In recent years, however, recyclable and reusable materials have, for the most part, been captured by local governments and private companies. Would landfills attract a crowd were they to be suddenly liberated? It is hard to say, but food waste, in particular, is being tackled in various schemes rolled out to tackle poverty. Some of the more exciting schemes share with the Uruguayan case the experimental potential of waste as decommodified substance. Real Junk Food cafes use food recovered from skips or donated by supermarkets and restaurants to make meals that are served up on a 'pay as you feel' basis, with customers encouraged to pay what they can afford or what they believe the meal is worth. The experimental model of payment is made possible by the fact that the restaurants do not have high outgoings for food, and ingredients can be recovered or donated, precisely because they have been decommodified, usually due to their having reached the end of their shelf-life and sell-by dates. Unlike schemes where donated food is distributed through charities and NGOs to a deserving poor, often on a means-tested basis, Real Junk Food cafes do not impose any criteria on access, encouraging attendance of people from a range of social backgrounds. Yet lost in both food distribution schemes and surplus cafes are elements of autonomous labour and the excitement of the treasure hunt found in waste-picking, whose closest parallel in the Global North can be found embodied in the activity of dumpster divers.

Conclusion

When cycling across Cambridge's expansive Midsummer Common or vegetable gardening at my allotment in neighbouring Little Shelford, I have often wondered whether English commons have only been saved from privatisation and development in places of privilege, where councils are not in such dire need of revenues or where residents are well-organised and well-connected enough to preserve them. Such green spaces certainly seem a world away from my fieldwork at Montevideo's Felipe Cardoso landfill. Yet it was in living next to the dump that I had the first opportunity in my life to plant a little vegetable patch, on adjacent land, inviting the kids who lived nearby to sow, care for, and harvest something of their own. Indeed, some of what we planted had been sourced from another type of commons – the *cantera* – such as the potato plants that Nona directed me to as we crossed the inactive landfill of Usina 5.

I had just started the vegetable patch, so we agreed to return, on horse and cart rather than bicycle, and pull out a few crops that could be replanted. Back then, I had little idea of what a potato plant looked like, so Nona signalled and I dug them out. Their roots gave an indication of what lay beneath, as embedded in several tubers were small pieces of plastic and polystyrene. Nona and I decided to plant them anyway, and we boosted their growth with the help of bags of free compost that I collected when visiting the municipal composting facility on the outskirts of the city. Might the potatoes have grown out of a load dumped at Usina 5? Every day at Felipe Cardoso, tonnes of surplus fruit and vegetables from Montevideo's central market are buried after clasificadores have salvaged what they can, hauling bags of ex-commodities onto their shoulders and then back home. How many potatoes and other tubers manage to survive, growing upwards in search of light as layer after layer of refuse was piled on top? When Usina 5 was closed, did the most tenacious crops finally reach the surface? Could the layering of waste upon potatoes constitute an extreme example of 'earthing over', where earth is piled upon potato plants to increase yield? Whatever their provenance, when harvested, they tasted just as good as those that I would later grow from mail order seedlings on the plot I rented from the local parish in Little Shelford.

In this chapter I have argued for expanding our definition of the commons into the uncommon territory of modern landfills, in a way that recognises not just the legacy of the original English commons as an inspiration for activists, but the very real continuities that can be found with how poor and vulnerable groups claim, practise, and defend access to urban resources today. Although commodification and exchange value are certainly more prominent at the *cantera* than in certain traditional rural commons, the conditions of contemporary capitalism make it difficult for this to be otherwise. In spaces like the *cantera*, use and exchange value, sharing and selling coexist – but in recognising this, we take nothing away from the importance of such messy commons for the poor, the injustice of hygienic enclosure, or the bravery of any resistance to it. The *cantera* also represents an autonomous space similar to that of the traditional commons; one with potentially greater earning potential than low-paid, unstable waged labour and thus an ally against precarity. What Millar (2015) calls the 'relational autonomy' of wage-less work at the landfill holds an attraction that cannot be explained by narratives of urban survival alone. This chapter's comparison with the historical English commons has hopefully demonstrated that, as well as spaces of risk, landfills can be sites where refuge can be sought and generosity flourish. The Felipe Cardoso waste-pickers might not be pure post-capitalist subjects and they certainly challenge Uruguayan ideas of infrastructural modernity in waste management. But without wishing a return to 'salvage ethnography' (see Clifford 1989), I would argue that there are many things, from materials to social relations, which are worth salvaging from the dump. The type of desirable ethical subjectivities and caring relations enabled by the flow and distribution of discards forms the topic of the following chapter.

3

Classifiers' Kinship and Embedded Waste

The work of clasificador families in Montevideo has often been considered problematic in local academic and policy circles. On the one hand, this is because the nuclear family – perhaps involving child labourers – is often taken to be the basic unit of production (IP & MA 2012: 1). On the other hand, it has been suggested that extended clasificador 'clans' and their territorial and kinship allegiances, have undermined attempts to cooperativise the sector (e.g. Sarachu and Texeira 2013: 8). Indeed, the latter is a concern present in the work of early anthropologists of cooperatives such as Peter Worsley (1971: 24), who, suggested that 'traditional ties of kinship and neighbourhood, caste and ethnicity, too, often work against the requirements of strict economic solidarity', and argued that 'established solidarities may be dysfunctional for the cooperative rather than a social foundation on which modern cooperation can be based'. Even further back, Hannah Arendt (1998 [1958]: 24) had argued that the very emergence of the political in ancient Athens rested on breaking the power of kinship over collective life and relegating it to the private sphere.

In my field site, I found neither the presence of child labour nor an incompatibility between kinship and cooperative forms: instead, kinship-based but diverse organisational models undergirded the waste-picker economy. This chapter argues that kinship is an important and indeed valuable mechanism through which the fruits of the waste commons are distributed. In the following pages, and through short case studies of three sites of clasificador activity – a cooperative, a family yard, and the landfill – I provide a detailed ethnographic account of the ways in which discards are commoned, appropriated as sources of value, and act to constitute relations of care between kin in and around the landfill. David Boarder-Giles (2015: 89) has written of two kinds of work that combine to make waste: that of discarding things and of 'subsequently refusing them recirculation in the contexts of their former "social lives"'. This chapter is concerned with the obverse process by way of which rubbish is undone through labours of 'unwasting'. In the context of a 'crisis of care' (Fraser 2016) in many parts of the world, where different pressures squeeze the social capacity for raising children, supporting family members, caring for friends, and holding households together, can the flow of discards sustain care between kin? Can the 'informal' waste trade be considered a space of hope as well as one of suffering and exploitation?

In the following sections, we meet different, interconnected families working the waste around Felipe Cardoso, and reconsider in the process the connection between kinship and poverty, care and commoning. Waste, in this account, is not merely hazardous, risky, or the excess of social relations of production: it also co-produces practices of precarious caregiving against the background of the historic exclusion of informal workers like clasificadores from the benefits of labour-based citizenship and social security (see Dimarco 2011). An exploration of kinship-based access to and distribution of waste nuances rather than undermines my commons framework, since it is the decommodification of waste and the position of clasificadores outside of wage labour that allows them to distribute materials and labour to vulnerable family members. The forms of caring that I discuss in this chapter are similar to those explored by Clara Han (2012): they are shaped by and face limits in situations of poverty and economic precariousness, including the difficulties of caring for relatives with problems of addiction. Like Han, I seek to challenge portrayals of urban poverty as characterised overwhelmingly by survival and competition, focusing instead on the weave of relational, care-making practices that occur within the sphere of intimate life, which in this case overspill into labour and its distribution.

As I have noted, most social science research on waste-pickers in Latin America has focused on – and been sympathetic to – the minority grouped in associations and cooperatives. These 'organised' waste-pickers have often been championed and contrasted to those working 'individually' in the informal sector, who are portrayed as individual and individualistic workers (Carenzo and Miguéz 2010). Indeed, anthropological literature on the Latin American urban poor at times reproduces a division where marginal groups are seen as either engaged in the forms of political life of the *polis* (e.g. Bryer 2010; Centner 2012) or reproduce forms of 'neoliberalism from below' (e.g. Auyero and Berti 2013; Gago 2017; Macedo 2012), a form of critique that arguably reproduces the Marxist division of the organised from the lumpenproletariat. The 'persistent life of kinship'

(McKinnon and Cannell 2013) among the Uruguayan popular classes troubles the division of waste-pickers and the urban poor more generally into those organised in political and social movements and atomised individuals. Within a single extended family or neighbourhood group we can in fact find quite different models of waste-work, which simultaneously sustain a material infrastructure of kinship care and provide the backbone of recycling chains that go far beyond the *cantera*.

Ethnography thus allows for a close examination of the complexities of capitalist supply chains and the diverse models that underpin them (Tsing 2015). In Narotsky and Besnier's (2014: S4–S5) terms, it can be a 'precious instrument that draws attention to the historical production of specificity and its role in structuring differentiation' within processes of capitalist accumulation. In order to avoid a narrow focus on families that disregards wider structural relations of inequality, this chapter draws attention to the origins of the waste that clasificadores come into contact with and, where relevant, the destination that it travels on to. The focus on Montevideo that characterises the following three case studies is also expanded upon in the subsequent section, where I discuss the distinctiveness of waste as a source of labour as compared with other instances where labour and resources are distributed along family lines and structure capitalist forms. Drawing a final comparison with Mao Mollona's (2005) work on the re-embedding of capitalist relations in Sheffield's now-fragmented steel industry, I point out that some informal recyclers have never been wage labourers and thus only recently has their labour come to be disembedded, with attendant risks for the maintenance of kinship care relations.

The waste-picker cooperative

In late March 2014, I was sat in the back of the Pedro Trastos Cooperative truck as we returned from the large Montevideo distribution plant of Anglo-Dutch multinational Unilever. The cooperative had a regular pick-up (*levante*) of the company's waste, and was one of the few third sector organisations to have accredited waste transporter status, which enabled them to take discards to Felipe Cardoso. The Unilever cargo consisted of damaged products and waste created in the distribution process, as goods mostly manufactured in neighbouring Brazil and Argentina were unpacked. The contents were doubly enclosed, first within rubbish sacks, then in the truck. Such waste was, according to regulatory guidelines, assimilable to the category of household rather than hazardous waste, but it still represented negative value for the multinational company, which had to pay the cooperative for the disposal service.

We didn't quite make it to Felipe Cardoso, however. Instead, we turned off a few streets before and passed recently built social housing complexes, a smallholding, some shacks, and a community centre run by Franciscan nuns. At the end of the street, we reached what appeared to be a cul-de-sac but then continued down a dirt path that opened out onto mixed terrain. To our left was a large aluminium shed and several self-built homes; to our right, green vegetation interspersed with discards and pigsties. It was a wasteland all right, but not the one we should have been arriving at. This was Trastos territory, where spaces of cooperative waste-labour mixed with family homes. It was where I had spent the summer of 2010 and cut my teeth as a clasificador, conducting fieldwork with the Trastos cooperative for my undergraduate research.

It had begun to rain when Nico stopped the engine of the truck. He was the president of the cooperative, a charismatic and enterprising figure whose previous occupations included shantytown bar-owner and recycling intermediary. He was from a large waste-picking family, however, so when his uncles and relatives working at the landfill began to get organised in 2002, he quickly rallied to their cause. After a fact-finding trip to visit waste-picker (*catador*) collectives in Brazil, Nico and other visitors became convinced that the cooperative model was the best way forward for Uruguayan clasificadores. His first cooperative experience was with COFECA, but he subsequently led a group of his uncles to form a smaller, breakaway group: the Cooperativa Pedro Trastos (named after another, deceased uncle). The cooperative included Morocho, Juan's stepfather, and Juan himself had worked for them as a driver. The name Trastos, meaning 'useless, old, or bothersome thing' (RAE 2020) in Spanish, had been given to Nico's grandfather due to his work in the waste trade, and was passed down to subsequent generations. While their legal surname was the patrician Alvear López, for all intents and purposes they were *Los Trastos*, one of the largest waste-picking families in the Cruz de Carrasco neighbourhood.

The cooperative had downsized somewhat since I conducted fieldwork with them in 2010: two members were in jail for drug-related offences, Nico's uncle Savia had suddenly passed away, and founding member Uncle Sordo had left following a dispute. So, on return from Unilever, only Nico's uncle (and Juan's stepfather) Morocho was there to help him open the truck's tailgate, soon joined by Uncle Kela who had wandered over when he heard us arrive. Kela was no longer part of the cooperative but he was still family, so no-one objected to him taking away the odd product to freshen up his kitchen or his armpits. Nico positioned himself at the back of the hold and began pushing full transparent rubbish sacks towards us. As we emptied them one by one, out tumbled a colourful assortment of cleaning products marketed by Unilever in Uruguay: black cans of Axe deodorant, packets of yellow Jif oven cleaner, bars of white Dove soap, bags of pink and blue Nevex washing up powder, and one-wash sachets of Sedal shampoo. The packaging of many of these products had been damaged during importation, causing flecks of white washing up powder and squirts of shampoo to mix with the falling raindrops and form a frothy lather when trampled underfoot.

What was wrong with the Unilever discards that caused them to be thrown away? They had not gone rotten, become toxic, or spoiled. Their chemical make-up had likely stayed the same and, judging by the floral aroma that drifted over to perfume the Trastos's pigsties, they hadn't lost any of their scent either. What had been damaged was what Reno (2015: 106) calls 'bundling waste' – packaging materials that 'help other things last ... display their durability to others [and] prepare the representational grounds for the commoditization of discrete and saleable things'. Unilever sold their products to commercial outlets in numbered bundles, meaning that if one small shampoo sachet in a roll of 30 burst, the remaining 29 were also converted to waste. I had been given a few months' supply of these by the Trastos in 2010, when they noticed that I had come into work with cement still in my hair from the previous day's labour on the building site and thought that I was in need of a wash. For single items too, packaging enabled transport and constituted a crucial component of exchange value. A cracked bottle disturbed the homogeneity and replicability of the Sedal Wheatgerm and Honey Formula Shampoo that Kela cheerfully carted off in a large tub, and Unilever had no use for substances that maintained their chemical integrity but not their plastic shells and commodity status.

Yet instead of heading for the dump, these substances were diverted towards acts of kinship maintenance (see Carsten 1995). In all, around half the load could be recovered and was mostly split between Morocho and Ainara, whose husband Nacho was a Pedro Trastos member imprisoned several months earlier for the possession of marijuana. He still figured on the cooperative's books, and his wage, along with a share of *requeche*, went to supporting Ainara and their four children, a form of informal social security. Ainara might use the products herself, relieving the family of a household expense. Or she might realise the latent 'commodity potential' (Appadurai 1986) by selling them to neighbours, receiving in return money that was used to buy bread, *yerba mate*, biscuits, and cigarettes that we brought her husband in prison, helping him to get through a spell behind bars. As with dumpster divers, discarded things 'could become many things ... could indeed be reintroduced into the market ... [but] may undergo a parallel kind of nonmarket economic circulation that obviates in some way the logic of the market' (Boarder-Giles 2014: 108).

Nico's decision to adopt the cooperative form, meanwhile, represented a pragmatic and timely alignment with hegemonic government waste-picker policy and an explicit attempt to intercede in capitalist recycling markets where waste-pickers, as the weakest link in the chain, received the lowest remuneration for their labour of waste classification. Cooperatives like Pedro Trastos were meant to cut out the middle-man or intermediary, allowing the cooperative to sell directly to industry, thus obtaining higher prices for their materials. What Nico established and could distribute to kin was an earlier interception of waste, and agreements with businesses, achieved in part through his charisma and social capital. The cooperative form also enabled them to access support from the state and resources from NGOs, such as a truck and a shed donated by the Canadian Embassy.

In her analysis of charity and social enterprise recycling initiatives in the UK, Catherine Alexander (2009: 239) points out the difficult position that such enterprises are placed in, where 'the third sector, freighted with the morality of embeddedness, kin relations, and households, is also burdened with the task of restoring social equilibrium on a larger scale whilst shackled to short-term funding and uneven labour supplies'. Different, though equally burdensome, pressures were placed upon cooperativisation as a proposed solution to the structural exploitation of Uruguayan waste-pickers. Like the post-Soviet Estonian miners studied by Keskula (2014), clasificadores were expected to move rapidly from a situation where the management and distribution of labour was embedded in family relations to one structured by a different form of economic tie. Yet, through careful and charismatic management, Nico was for a while able to embed wider kinship relations in the cooperative, creating a hybrid that leveraged his authority in the family to ensure the disciplined labour force necessary for the regular collection of discards from large formal sector companies such as Unilever. Still, while he accepted these benefits of formalisation and cooperativisation, Nico did not renounce the clandestine resource also enjoyed by his Aunt Natalia, whose family worked on the same contiguous piece of occupied land.

The family yard

When the path at the end of the Trastos cooperative terrain becomes too narrow for a car or truck, it is possible to continue on foot through shrub and light forest, reaching a settlement adjacent to the old Usina 5. This was the home of another of Nico's uncles, the recently deceased Marco Trastos, whose family continued to live and work on the site. Marco and his wife Natalia had divided the space into a domestic sphere and a much larger area where they could receive, classify, and landfill discards. In the last years of his life, Marco was mostly to be found sipping sweet red wine from a plastic bottle at the entrance to their home, and that was where I met him back in 2010, when I went to collect wages from his daughter Olivia, who had contracted me to work as a labourer on her cooperative home while she looked after her children. His wife, the converted evangelical and now teetotal Natalia, took charge of the family business, employing her children Carlos, Juancho, Olivia, Sara, Cachula, Nono, Nona, and Mara in the task of classifying waste. I came to know the family well because many of the siblings became our neighbours in COVIFU. Unlike the Cooperativa Pedro Trastos, Marco and Natalia's business was fully 'informal'. Pedro's family didn't own a truck, so instead accessed waste by way of a division of labour with another neighbourhood family. The Motas collected daily from the upmarket supermarket chain Grands Magasins in their truck, appropriated the most valuable, least labour-intensive materials, and then dropped off the rest at Natalia's.

The expansion of the supermarket model in Uruguay took place from the 1960s onwards. Of the current existing supermarket chains Grands Magasins is the oldest, founded by a British immigrant in 1866 with a single shop in the City's Old Town (Ciudad Vieja). It wasn't until the 1960s that Grands Magasins expanded to include several stores, and other supermarket chains such as Disco and Devoto were founded. One Afro-Uruguayan informant, Omar, grew up in the same neighbourhood as the Devoto brothers, and worked in one of the chain's first stores in Malvín, where he remembers the introduction of plastic bags in the 1970s. As Néstor, another older informant told me, 'When I was growing up everything was sold by the kilo and wrapped in paper in corner shops (*almacenes*), like flour for example.' 'Then came the supermarkets (*super-*

mercadismo), with their packaging and bags and that was the end of that,' his wife Norma added. Plastics and supermarkets in Uruguay thus went hand in hand in a symbiotic relationship that eventually would provide fodder for the recycling industry, of which Natalia's family provided an important link in the chain.

Yet unlike in the UK and the United States, there is still a strong culture of neighbourhood fresh food markets in Uruguay, known as *ferias*. Thus, in Flor de Maroñas, we joined our neighbours in carrying out a weekly shop for fresh fruit and vegetables at the Sunday *feria*, to be complemented with *requeche* from the *cantera* that could be procured during the week. The *ferias* generally maintain their own customs of waste disposal, where left-over or low-quality fruit and vegetables would be left out in boxes on the pavement at the end of the day, an easily accessible commons for regulars and passers-by. This can be contrasted with supermarket waste, which is not only stored away in bins to be transported in trucks but also involves a much greater mixing of waste that affects quality and challenges the possibility of recovery. *Feria* waste is carefully laid out according to market stall: fish, chicken, fruit and vegetables all have their place. The supermarket waste that made it to Natalia's yard, on the other hand, came churned together in trucks, in a temporal process of wasting that made clasificador unwasting that little bit harder.

Effectively, Natalia's family yard acted as a site of flexible resource-extraction for the siblings, who were paid US$30 by their mother for a day's work recovering plastics, paper, metals, and whatever they wanted to take home for domestic consumption. Even those who did not work at the site regularly were able to rely on it as a source of labour. Carlos, a foul-mouthed loveable rogue who walked with a jerking limp, was a heavy drinker who also suffered problems of addiction to *pasta base*, causing his appearance at work to be erratic. Olivia, who by her own admission easily became bored with an activity, dropped in and out of work there during my fieldwork period. Sara's husband Esteban preferred her to stay at home to look after the house and kids but she worked at Natalia's for a few months to earn money to pay for her eldest daughter's fifteenth birthday celebrations. Cachula, the youngest daughter, mostly cared for her two infants, but turned to her mother's yard while her partner was out of work. That this was a space of care rather than exploitation was recognised even by Montevideo's then Mayor Ana Olivera, who had visited the site after a neighbour complained about the smoke resulting from the family's burning of rubbish. According to the Trastos who were present, and much to the annoyance of the municipal waste official who accompanied her, Olivera conceded that she did not see a place of child labour but one where women were working hard to support their families.

It was through classifying waste at Natalia's yard towards the end of my fieldwork period that I came to understand how the constant flow of materials helped to sustain bonds of kinship. I had considered working with Natalia for a while but was put off by her reputedly fierce temperament. Still, when we met at her home – whose uncluttered interior of evangelical imagery contrasted with the cornucopia of waste materials that surrounded it – she seemed happy with the offer of another pair of hands that would get dirty without requiring payment in return. My credentials had also been boosted by my appearance at a few church services. 'People outside of the family don't seem to like it or stay very long', Natalia warned me, and I soon realised why. I was hardly unaccustomed to waste-work, either in Uruguay or in Scotland, but the supermarket waste Natalia received consisted principally of packaging, damaged products, and spoiled food. The small amounts of plastic, cardboard, metal, and paper that we sought out were thus mixed together with rotten loaves of bread, mouldy grated carrots, and foul cocktails that stained our clothes. One valuable material that appeared regularly consisted of bloody, transparent plastic meat bags, which attracted flies and maggots that Nono and I wiped from our arms in the summer heat. These were trucks that clasificadores at the *cantera* would likely have ignored.

Compared to the supermarket food waste that I had recovered in St Andrews, this was thus of lesser quality. Perhaps this was because supermarkets in Uruguay threw less away, or perhaps better-quality food had already been scavenged further upstream, just as the Motas skimmed off the more valuable recyclables. Even the Grands Magasins loads contained the possibility of an exciting chance discovery, however. On occasion, I would find luxury items not sold in the local supermarket: a burst bag of cashew nuts, an unopened packet of Lavazza coffee, an imported German lager. Natalia's family and I alike delighted in the pleasures of one of these chance finds. Alongside these exceptions, there was the regular *requeche* of mixed vegetables that Natalia transformed into lunchtime stew or pigswill. There was no need to bring a snack either, as I followed Carlos's lead in tearing hunks of bread from salvageable loaves and improvising sandwich fillings with the rind of different cheeses or the ends of ham loins that arrived from the supermarket's delicatessen counter, the only time during the course of my fieldwork that I would eat serrano ham or manchego cheese.

Then there were objects of less immediate consumption. Reno (2009: 35, 2015) has written of how garbage workers at the Michigan landfill where he conducted fieldwork assembled discards into 'masculine projects' such as building cars. Nono was also something of a self-taught mechanic who loved fixing and restoring things. You don't find too many car parts in supermarket waste but we encountered plenty of broken toys from their children's section. One day, I watched him pull out several electronic toy dinosaurs, each with a particular defect. Over the next few days, he began putting them together – the sound box from one, the arm from another – assembling a figurine to give to his six-year old son. Nono didn't earn enough money to be able to buy new toys from Grands Magasins but with these efforts he could live up to the role of caring and crafting father. Olivia, meanwhile, made sure that her daughter's fourth birthday was the envy of the neighbourhood by saving up for it by working at the yard, and fashioning decorations from coloured paper recovered from the trash. The supposed absence of economic value brought materials into the waste-stream, while the recognition and recovery of its latency enabled the constitution of particular values, such as responsible parenthood (see Graeber 2001).

Despite my focus here on the positive aspects of such material encounters, the nature of the Grands Magasins arrangement and the composition of their waste-stream indicated that the risky and exploitative side of waste (work) was also present. The supermarket and its millionaire owner, a descendant of its British founder, paid very little for the collection of their waste and were spared some of the costs of disposal at Felipe Cardoso because much of it came to rest on the land that Natalia's family occupied. Over the years, the lake that Nono remembered swimming in as a child had been polluted by accumulating rubbish and pigs kept by the Trastos that had fallen in and drowned. Interestingly, Natalia and her family argued that they were performing a service by levelling the land with waste, a waste disposal method previously advocated then discarded by the state. As Penelope Harvey suggests then, past ways of dealing with leftovers also 'generate their own leftovers, not just all that is most toxic or corrosive but also other ways of defining the problem that open the way to solutions at a different scale' (2013: 70). Ideas, as well as material processes of disposal, leave behind their own residues.

Both before and after Natalia's family in the waste and recycling chain were actors who boasted greater capital, income, and status. On one side were the Motas, who were paid for dropping off the waste but barely got their hands dirty and were often seen whizzing around the neighbourhood on expensive motorbikes. On the other were the recycling intermediaries who bought materials already neatly classified and sold them on for a tidy profit. The situation of Natalia and her family is that which most closely fits a depiction of 'super-exploitation' at the bottom of the global recycling ladder. Subordinate to another neighbourhood family that appropriates the waste requiring least labour to transform it into a commodity (e.g. pre-classified cardboard), both families in fact worked for a third, the English-descended supermarket owners.

It is not only, then, that waste has a dual nature, possessing both a 'potential to turn into money' and a 'tendency to toxic decomposition' (Harvey 2013: 67): access to different wastes at different points in the waste-recycling chain is also unequally distributed, while waste's structured flows can expose actors to social risks and stigmas that go beyond the materiality of waste itself. Such a realisation warns against a purely celebratory narrative of human care built out of discards. In other ways, however, Natalia's model was relatively benign and advantageous for her children, who could vary the number of days they worked in a week, month, or year. Whereas the landfill was a difficult place to work for women, who were exposed to the humour and advances of men, at Natalia's yard they could work as a closed family, supervising their children and exposed to non-kin men only through contact with intermediaries.

The landfill brothers

Natalia's children had all been raised on the site, and the Azucarero brothers were never too far away. Sara and Esteban, and Martín 'Azucarero', were all now in their early thirties and neighbours in COVIFU Rural but they didn't speak to each other during most of my year's fieldwork. When I interviewed Sara and Esteban in the last weeks of my stay, I asked about the circumstances of their joint upbringing:

> *Sara*: I know them all, we were brought up together. From the age of nine we used to go to the *cantera*. They were all brought up at my mother's house: Mariposa [Butterfly], Martín, el Gallego [the Galician], la Momia [the Mummy], el Pegado [Glue] ...

Patrick: They were brought up at your house?

Sara: They would stay at my house, they were called the *guachos* [orphans, abandoned kids], those who wouldn't stay in their parents' houses.

Patrick: You told me that your parents used to drink a lot, but they still had the time to bring up other kids?

Esteban: They were brought up like pigs … 'go and look for some *requeche* in the *cantera*!' And they all ate what they liked …

Sara: Don't lie! My father used to make pots of food. Be quiet, boot-licker [*lambeta*].

Esteban: Yeah, from *requeche* and everything that came out of the rubbish …

Sara: And? So what? What's wrong with that?

The conversation highlights conflicting opinions about caring-through-waste. Esteban's father was a plumber, so although he grew up near the landfill and went to school with kids who hung out there, he was a step removed from the world of waste-picking. For him, the care Sara's parents provided for the Azucarero boys was risible: it was really the *madre cantera* who provided them with sustenance. Sara, on the other hand, lauded her parents for distributing *requeche* beyond their offspring to unruly neighbourhood *guachos*, who at least got fed and had some adult oversight. The Azucarero boys were a handful, even by their own accounts. Martín told me that he only went to school to steal other children's pocket money, while his brother Michael, another neighbour, was kept home permanently after a violent incident. I introduce them here because as they grew up, the Azucarero brothers also used access to waste to sustain relations of kinship care. While Nico did so by adopting a promising cooperative form being promoted by the state, and Natalia distributed access to waste in the informal sector as she had done for years, the Azucareros had yet another arrangement. To find them, we need to continue our journey from María's yard along Camino Oncativo to the active Usina 8 where they worked.

Like Trastos, 'Azucarero' was a family nickname born of an association with a particular material, in this case the sugar cane (*caña de azucar*) that the boys' mother cut down from the surrounding commons and mixed with mud to build the family's first home in the Cruz de Carrasco. Even among waste-pickers, these building materials were distinctive enough to merit a sobriquet. They indexed poverty and resourcefulness and were considered rustic in comparison with the metal and wood that most clasificadores scavenged from the landfill and surrounding area to build their shacks. When I arrived to conduct fieldwork, the next generation Azucareros lived in cooperative 'material houses' (*casas de material*). They might still have been waste-pickers, but they were successful ones, enjoying exclusive access to the landfill in the afternoon. This right had been earned through another affective relationship mediated by waste (work): the friendship between the Azucarero brothers and municipal landfill workers.

Such relations began when the eldest son, Mariposa, used to gather *requeche* to support his mother as a boy. This was a time when, both clasificadores and landfill employees informed me, municipal workers would commonly appropriate the contents of dump trucks in order to supplement their meagre salaries. Mariposa told me that he spent time at the smallholding of one municipal worker, Molina, separating the contents of trucks in exchange for a share:

> They dumped trucks from the markets at the farm and we used to separate the fruit and vegetables for the animals. That's how I made friends with him: I worked for him, as they say. I got seven or eight boxes to take home to my mother and the rest was for the animals. It was a mutually beneficial agreement.

The labour–friendship nexus took on different iterations as Mariposa grew older and Molina's son became landfill foreman. For a while, he assisted with the ostensibly municipal task of *atracando*: rapidly directing the trucks to where they should dump their materials ('It's a job that needs real skill to avoid a queue of traffic, a skill that I had but many *municipales* didn't'), recalling the work that Danny used to perform for the council at the local amenity tip in Cumbernauld. During my fieldwork period, Mariposa worked as a waste-picker but still performed a municipal function in restricting the number of waste-pickers who could enter in the afternoon to his siblings ('Ask the foreman if he's ever had any problems with us in the afternoon, ask him …'). Mariposa was clearly proud of his friendship with the *municipales* ('I've been to their houses to eat, they've come to mine') and the sentiment was obviously reciprocal, for when I interviewed the landfill foreman and told him that I was living in COVIFU, his eyes lit up: 'I've a friend who lives there: Martín Azucarero!' Friendship with the foreman paid dividends for Martín the day that he was offered the opportunity to receive and classify several valuable trucks that were being diverted from over-stretched recycling plants on a daily basis.

Through their access to waste, Martín and Mariposa managed to combine individual progress with the care of more vulnerable siblings. The younger Azucarero brothers – Michael, Gallego, and Pegado – had all suffered problems of addiction to *pasta base*, likely an outcome of their coming of age at a time when the drug started to destroy the lives of young men in Montevideo's poorer neighbourhoods (Suárez et al. 2014). Of the brothers, I knew Michael and his life story best: the homelessness, the overpowering addiction, the multiple attempts at rehabilitation, the relapses, his new family and life at the Aries recycling plant. During all his tribulations, he told me, Martín was 'always there for him', either lending him money that he had earned through selling recyclables, or facilitating his access to work at Felipe Cardoso, while Mariposa did the same for Gallego and Pegado.

Embedded waste

The density and multiplicity of clasificador kinship relations described in this chapter suggest that asserting the parent–child dyad to be the key relationship of care, or indeed exploitation, is mistaken. Dimensions of social life that have been recognised in regional scholarship on low-income neighbourhoods (e.g. Fonseca 2000), such as weak conjugal bonds, large family sizes, shifting father-figures, and the practice of taking in abandoned children (*hijos de corazón*) meant that the make-up of families could be extremely heterogeneous. Nico, for example, was older than his Uncle Nacho, who was not a biological relation but had been raised as a Trastos after running away from home and being taken in by Nico's grandmother, an example both of 'kinning' (Howell 2003) and of how care itself can create kinship in the absence of biological connections (Drotbohm and Alber 2015: 8). The Azucarero siblings, meanwhile, had different fathers, and older brothers like Martín and Mariposa often took on the principal caregiving role for younger siblings.

The 'unclear families' (Simpson 1994) of Uruguayan waste-pickers complicate narratives of generational social reproduction dwelt upon in policy accounts and attempts to intervene in the reproduction of poverty. The generational link is clearer in the work of Minh Nguyen (2018) on Vietnamese waste traders, where parents move to the city to work in the waste trade, leaving their children with grandparents in rural villages, reconnecting and cohabiting with the children only when they reach adolescence. In that case, economic structures and opportunities play a key role in reshaping expectations of appropriate gendered kinship caring roles, where grandparents and men more generally often become the primary caregivers during infancy and early childhood. This chapter has focused more on what maintains kinship over the life-course than what creates it in infancy and the different family cases cited here indicate the ways in which discards hold clasificador families together rather than pulling them apart.

My informants distributed access to waste based on pre-existing kinship bonds (uncle, brother, child) but it was also through waste that practices of caregiving between adult family members were enacted. Kinship was not only biological connection or mutual upbringing but the stitching together of a patchwork safety net from the offcuts of industry and commerce. Through distributing waste and waste-work, Nico, Natalia, and Martín embodied the figures of the caring uncle, mother, and elder brother, enabling the 'fulfilment of an expected form of behaviour associated with a specific kinship role' (Drotbohm and Alber 2015: 7). Amidst a rush to recognise emergent new subjectivities associated with waste and contamination – for example, Hawkins's (2006) 'anxious recycler' or Hecht's (2012) 'being nuclear' – we should not forget that waste-work also sustains more traditional subjectivities such as those expressed through idioms of kinship.

Back in St Andrews, the capture of surplus food from supermarket bins also helped to build friendships and communities. At Watson Avenue, where a group of my fellow student activists and dumpster divers resided, scavenged food enabled shared meals for the best part of two years. More than that though, much of the scavenged fruit and vegetables – far too much for a single household – was shared with the University Vegetarian Society, which provided budget lunches for thousands of students while our activities lasted. Just as in Uruguay, the recipients of a *requeche* meal didn't always know the provenance of their supper and, like clasificadores, we took a guilty pleasure in our secret, often revealed after diners had complemented us on a fine daal or cauliflower pakora. Without the 'skipped' food, it simply would not have been possible to provide such an affordable lunch for the Fife university's hungry vegetarian masses. The meals at Watson Avenue would also likely have been somewhat less nutritious, and the comradely bonds formed skipping, cooking, eating and protesting together would have been that bit less tight.

What the Uruguayan cases of waste-labour discussed in this chapter share is that they all connect practices of kinship and care with national and regional recycling chains. The breaking apart of the Fordist model and increasingly diverse forms of subcontracting and supply chains have undermined the extent to which capitalism can be seen as disembedding economy, economic relations, or 'fictitious commodities' like labour and land (see Hann and Hart 2009; Polanyi 2001 [1944]) from the dense web of cultural, customary, and moral relations in which they were and are entangled. As Palomera and Vetta (2016: 423) argue, 'capital accumulation is structurally inscribed in the everyday dynamics of social reproduction'. Two interrelated points have emerged from these discussions. First, the extent to which capitalism relies on non-economic bonds of kinship, affect, and communitarianism in order to extract market value and enable accumulation. Second, how this represents a concomitant cause for hope in its detractors since capitalism is less monolithic and all-encompassing than we might otherwise imagine, and forms of interaction that are not 'capitalocentric' are multi-fold and ever-present, if only we stay alert to them.

The particularity of the distribution of waste as a source of labour and livelihood, and this chapter's focus on care as labour *and* kinship drives my contribution to this debate. Let us briefly compare three other instances where labour and resources are redistributed along family and kinship lines in order to bring out what is unique about waste. First, there are capitalist family firms (e.g. Yanagisako 2002), where what is primarily passed down generations is capital and the means of production. Another, more traditional area of anthropological scholarship has been small craft/artisan workshops: there, skills, workshop style, and a commercial market are inherited (e.g. Cant 2019). Finally, we have cases where what is distributed is not primarily capital, means of production, or skills, but rather access to wage labour. This was historically the case of Argentine YPF oil workers (Shever 2012) and the large Indian steel firm Tata studied by Andrew Sanchez (2012). Clasificadores have no capital nor formal sector waged labour to pass on to their siblings and children. Instead, they pass on access to materials that can, through labours of classification and unwasting, be transformed back into commodities, consumed at home, or circulated as gifts through networks of friends and kin. Given their material work with metals, plastics, and paper, however, they can, like Argentine oil workers, provide a 'striking instance of the relevance of kinship to global circuits of capitalism' (Shever 2012: 88).

In the cases described, waste-work is embedded and essentially undisciplined labour from the perspective of rationalised capitalism, and it can afford to be precisely because of the materiality and status of waste. Unlike a harvest (where vegetables might rot), a commercial business (where customers might go elsewhere), or indeed any other formal business (bound by contractual obligations), the consequences of not turning up to process waste are incredibly small. Who would notice or complain if an ex-commodity remained an ex-commodity, if waste ... went to waste? Access to waste thus enables forms of work and labour that are compatible with, but also help to alleviate, forms of precarity. In all three cases examined here – the cooperative, the family yard, the landfill brothers – some of the kin to whom waste-labour was distributed suffered from problems of addiction to *pasta base*. During their initial stages of recovery, none of the users would have been able to hold down regular forms of wage labour for the state or private sector. The forms of flexible working – most salient in the case of the landfill, least in that of the cooperative – were simply not available elsewhere, and these were enabled by the fragmented availability of the waste commons.

Unlike the Bolivian garment workers in Buenos Aires studied by Gago (2017), I did not encounter exploitation *between* members of the waste-picker families studied. Yet we are left with the question of the exploitation of these families by intermediaries and more formalised actors in recycling supply chains. A comparison with the Sheffield steel workers studied by Mollona (2005) can be instructive here, where he argues that there was a 're-embedding' of labour as steel companies privatised, fractured, and turned to subcontracted labour, forcing workers to rely on a mix of informal and formal strategies to get by. Thus, he concludes:

> the blurring of the times and spaces of reproduction and the re-embeddedness of economy into society ... translates into a transfer of the organizational and welfare costs of production from the capitalists onto the workers themselves, rather than in an improvement in their living conditions. (Mollona 2005: 196)

Waste-picker labour in the families I worked with, in contrast, had never been disembedded in the first place: many have never been wage labourers and comprehensive forms of care and welfare have never been provided by either the state or paternalistic companies. Just as waste-picker leader La Pato says that clasificadores have had to 'invent their own work', they have also had to invent their own forms of care, partly achieved through the distribution of waste-labour.

We must also ask, with regard to the forms of exploitation present in current recycling chains that rely on the social infrastructure of kinship labour, what are the alternatives? One is the development of the cooperative form, which has been relatively successful for Pedro Trastos but has largely failed for other Montevidean waste-picker collectives. Another is the establishment of municipal recycling plants that will be explored in the following chapter. Yet all too often the removal of waste as a source of labour that clasificadores are able to distribute to family members in precarious situations heightens the precarity and poverty of such families, disembeds and alienates labour, and removes one of the few precious resources that clasificadores can use to establish themselves as kin who care. In recent years, this has happened to Natalia, as part of the land that she squatted has re-emerged as a commodity, successfully reclaimed through the courts by the son and heir of the deceased owner. While one of her daughters has obtained work as a cleaner while her children are at school, most of her children have curtailed their work, while her son Nono went on to work for the Motas, for a similar wage to that he received from his mother, but now uprooted from his family.

Conclusion

As Gabriela Vargas-Cetina notes, early studies assumed that 'cooperatives were modern forms of organization that superseded or were to supersede, in the long run, more "primary" forms of association based on the family, age groups, kinship or tribes' (2005: 231). Rather than representing a narrative of succession, this chapter has shown how cooperative and kinship relations are intertwined in my field site. I have also gone beyond loose notions of clasificador clans and accusations of child labour to elucidate how acts of precarious caretaking between siblings, which begin in childhood, continue into adult life and become entangled with the provision of waste (-labour). While kinship-embedded waste-labour sustains forms of exploitation inherent in capitalist supply chains, it is also co-productive of particular patterns of adult sibling caretaking. Such patterns relied on my informants' status as a non-waged poor with privileged access to the value and labour possibilities that inhere *en potentia* in the surplus materials of industrial production.

Lest my focus on caregiving be accused of painting an overly positive portrait of clasificador family life, I should underline that kinship, among my informants as elsewhere, can involve 'differentiation, hierarchy, exclusion and abuse' (Carsten 2013: 247), as well as 'promises and breaches of promise, acts and violations of intimacy, and acts of forgiveness and revenge' (Lambek 2011: 4). It is not that the Trastos were always caring for each other: there were clear material, social, and psychological limits to mutual assistance. But what appeared limitless was the arrival of Grands Magasins' waste, meaning that there was always work to be done, and a pot of *requeche* stew on the table. If they ran out of money, Natalia's children could always be sure of a day's work at her yard – no small help given that both Olivia and Mara were single mothers and Nona had to care for six children alone while her husband was in jail. Wounds between the siblings could be healed, and flared tempers calmed, while picking through mounds of trash or hoisting *bolsones* of recyclable materials onto each other's shoulders.

Whatever the unpleasant sensorial dimensions of the work, its rhythms were leisurely and even therapeutic, as we hid away conversing behind mounds of trash. 'I've worked at the landfill and COFECA', Olivia told me, 'but there's nothing like working at your mum's, with your family all around.' Whether through the dinosaurs Nono repaired for his son or the ornamental Chinese warrior that Sara glued together and used to symbolically guard her property, the siblings were not involved in 'waste-making labour' (Boarder-Giles 2016) but a labour of unwasting, through which they reconstituted things and simultaneously constituted themselves as subjects who cared – for each other, for their children, and for their homes. The appropriation of waste and its recirculation to recycling plants entails a reconfiguration of care that will be explored in the next chapter, where clasificadores are reclassified as 'vulnerable workers'.

4

Care, (Mis)Classification, and Containment at the Aries Recycling Plant

> Tell that old woman that I'm happy shitting in the woods.
>
> (Martín Azucarero, on being told of the new sanitary facilities in the Ley de Envases plants being promised by Montevideo Mayor Ana Olivera)

> It was a complete change. One always has to progress and not always be stuck doing the same thing. I'm happy at the plant.
>
> (Ana Clara, Planta Aries – formerly COFECA – worker)

Thus far, I have described how clasificadores' value recovery practices ran parallel and in opposition to a municipal risk-based approach to waste. At the end of the twentieth century and beginning of the twenty-first, the Intendencia's approach to clasificadores and waste began to change. Here, I explore these dynamics through a focus on care, (mis)classification, and enclosure at the Aries recycling plant, where many of my COVIFU neighbours were recruited to work in a public–private initiative that attempted to bring both workers and materials into the formal sector. The chapter sketches out how the provision of uniformed, formal employment for waste-pickers in recycling plants became an integral part of Uruguayan infrastructural modernity in the waste management sector. This occasioned a series of tensions with clasificadores, who sought to defend practices such as the recovery of *requeche*, the use of horse-and-cart transport, and heavy lifting and masculine behaviours associated with the *cantera*. Rather than being limited to Montevideo, the changes that I focus on are part of wider global circulations of ideas that can be traced back to, among other events, the 1992 Rio Earth Summit, and can be compared to historical events, such as the division of London's riverine poor in the 1700s, as described by Peter Linebaugh (2003 [1991]).

I begin the first section on care by tracing a brief genealogy of the change in municipal waste-picker policy, highlighting the influence of Uruguayan priest Padre Cacho over municipal policy makers as the elimination of 'rummagers' made way for the social inclusion of 'classifiers'. I suggest that in prioritising jobs for waste-pickers in recycling plants, authorities maintained the link between waste and vulnerability inherent to my conceptualisation of the waste commons but codified and gendered the criterion of vulnerability in ways that proved challenging for male subjects in particular, whose skills were devalued and bodies exposed as frail. In order to be cared for by the new paternalism of the Uruguayan state, clasificadores were obliged to become entangled in a citizenship project and embrace a particular process of subjectivation.

The second section turns from *homo vulnerabilis* to *homo oeconomicus*, arguing that institutional authorities misclassified both the workers and the materials that would enter the Ley de Envases plants. Confusing clasificadores for the extreme poor, authorities imagined that they would be happy with a minimum wage, disregarding their heterogeneity and the composite incomes available to them in the informal sector. The payment of what I term a 'waste wage' – non-pecuniary recognition of their environmental service and sacrifice – proved insufficient, particularly given the ban on the recovery of *requeche*, a move that I compare to the interdiction of dockers' recovery of wooden 'chips' in seventeenth-century London. Given the over-valuing of the domestic recycling fraction that entered the plants, workers had to turn to informal sector practices such as the sale of *requeche* in order to supplement their meagre salaries, an example of how informal economic activity continues to subsidise and even underwrite formalisation projects (O'Hare 2020).

The final section on enclosure draws on a translation of the Ley de Envases as the 'container law'. The law entails, I suggest, certain forms of continuity with the hygienic enclosure of the past, even if municipal waste containers were reprogrammed to protect property from theft instead of the population from harm. Alongside the attempted containment of materials and workers within the formal sector, the plants entailed an attempt to contain the excessive masculinity of workers, with vulgar and violent behaviour discouraged in ways that resonate with Bakhtin (1984 [1965]) and Mbembe's (1992) discussion of the 'aesthetics of vulgarity'. Popular forms of rowdy and carnivalesque behaviour were not entirely prohibited, but there was an attempt to keep these away from the conveyor belt and contained within proscribed spaces for licence, such as 'social tourism' trips to the countryside.

Both the idea of the 'informal sector' and its celebration (e.g. de Soto 2002), have come in for a sizeable amount of criticism in recent years, even if the concept has been largely reified in policy circles. Keith Hart (1973), who coined the term to describe the activity of young men in Accra, laments its transformation into a 'jargon word' and the ensuing loss of analytical precision (Hart 2008). There is a current intellectual fashion to address instead the question of precarity, precarious labour, and the precariat (e.g. Standing 2011), although this concept has its own problems (Breman 2013a). With regard to Uruguay, Fernández (2010, 2012) avoids using a formal–informal binary to separate clasificadores from official waste management actors, opting instead for the terms 'spontaneous' and 'institutional'. Municipal waste management is not entirely formal, she argues, because the Intendencia cannot guarantee that all collected waste stays in the formal sector. Clasificador activity is spontaneous instead of informal, meanwhile, because it is self-emergent and because, like elsewhere in the Global South, there is not much of a formal recycling sector to which it can be compared (Fernández 2012: 2).

Yet my research, unlike that of Fernández, deals with an explicit attempt to formalise part of the waste recovery and recycling chain, and transform workers from cooperativists in a workplace of flexible labour hours and cash-in-hand payments to employees in one of fixed working hours and formal wage deductions/contributions. In a key change since Keith Hart's original conceptualisation, the informal/formal binary is now also an ethnographic term that one encounters in the field when studying the implementation of labour policy across the globe, one that is explicit in policy documents and influential in shaping people's orientation to life and labour. In this chapter, I accept a conventional definition of the informal economy as economic activity for which taxes are not paid and social security benefits are not contributed to or accrued. This understanding fits with the commonality identified by Chen (2012: 488) in an otherwise diverse informal 'sector' globally, where 'they [all] operate outside of the reach of state enumeration, regulation and protection'. It also corresponds to Guha-Khasnobis et al.'s (2006: 7) restriction of the 'formal–informal continuum' to the question of the 'relatively high and relatively low levels of the reach of official governance mechanisms', avoiding both value judgements on the benefits of formalisation and the association of 'informal' with 'unstructured' and 'chaotic'. Finally, it builds on Lazar's (2012: 16) recognition that, despite criticism, the formal/informal binary remains a 'productive heuristic tool'.

Care

The Ley de Gestión de Envases y Residuos de Envases was first approved by the Uruguayan parliament in 2003 but only began to be implemented in Montevideo in 2014. The law seeks to raise a levy from companies that release unreturnable packaging into the economy and environment; to recover and recycle such packaging; and to bring clasificadores into the formal sector. Designed as a measure of corporate social responsibility, the law does not involve the direct taxation of businesses but instead relies upon voluntary contributions that producers and importers of packaging waste make to the Uruguayan Chamber of Commerce (CIU). The CIU then pays for plant machinery and workers' wages; the Intendencia coordinates the supply of waste material; the National Environment Division (DINAMA) approves waste treatment; and MIDES deals with clasificadores and the contracting of diverse NGOs to manage the plants. Four such plants were built in Montevideo in 2014 to employ 128 workers and I conducted participant observation at one of them, which I will call the Planta Aries. This plant was built principally to house workers from COFECA, the landfill-based cooperative we met in the introduction that was simultaneously disbanded as part of a shift in municipal policy away from clasificador cooperatives and towards NGO-managed plants. We can think of the plants as sustaining some of the characteristics of the commons, as they maintained the link between vulnerability and access to the waste-stream as a source of labour. Work at plants was classified as 'protected' and access to jobs based on a codified criterion of social vulnerability that gave priority to those who had already worked in the informal waste trade.

A genealogy of the Ley de Envases indicates that the roots of the ideology of caring for and accompanying clasificadores as vulnerable workers can be found in Uruguayan Catholic social work. During the dictatorship, Padre Cacho was a Catholic radical operating under the protection of Montevideo's archbishop, the quiet but virtuous Carlos Parteli (Clara 2012). He lived in a poor community in provincial Salto before returning to Montevideo in 1978, where he established himself first in the shantytown of Plácido Ellauri, then in neighbouring Aparicio Saravia, living in shacks (*ranchos*) no different from those of other residents. He wanted to move to the shantytowns, Cacho told a fellow priest, because 'that is where God is, and I want to find him' (Clara 2012: 29). Cacho did not only privilege the poor however, he also prioritised a subsection within them: waste-pickers, becoming known as 'the priest of the little carts' (*el cura de los carritos*).

The Catholic social workers around Cacho were influential in the Frente Amplio, and played a role in reshaping municipal social policy when the party's candidate, two-time Uruguayan President Tabaré Vásquez, was voted mayor of Montevideo in 1990.[1] A cross-party group was set up to deal with the question of informal sector waste-work and, according to founding member María del Carmen, whom I first met at a meeting of clasificador assistants (*técnicos*) in 2010, Cacho was consulted on its nomenclature. He assembled waste-pickers in the Intendencia to discuss the matter and the term 'clasificadores' was approved by a large majority. It was not the first time the name had been used, with a 1968 newspaper article making reference to a Comité Provisorio de Clasificadores de Basura (Provisional Committee of Rubbish Classifiers) that complained to the Intendencia of being denied access to waste, and thus to the exercise of their livelihood (*El Debate* 1968). In adopting the title for their Clasificador Working Group (Grupo de Trabajo con Clasificadores or GTC), María del Carmen told me that institutional sympathisers wanted to 'recognise their [waste-pickers'] role as a positive link in productive processes'. The rebranding of Uruguayan waste-pickers as 'classifiers' rather than 'rummagers' can be compared with the Cairene 'Zabaleen', which means 'garbage people' in Arabic and has been rejected by waste-pickers there in favour of terms that focus on their role in maintaining cleanliness (Furniss 2017: 31).

Cacho had particular influence over Tabaré Vásquez's director of public works, Martín Ponce de León. According to María del Carmen, 'Ponce' had brothers in the priesthood and 'a sort of personal, familial and ethical debt with Cacho'. Then director of sanitation, Carlos Paz was also supportive, with María del Carmen considering him a 'pioneer' in 'understanding waste in its social, cultural, and economic complexity', an example of the global but often vernacularised 'integrated waste management' model (see Harvey 2013: 63; Sorroche 2015). Ponce's time in office was remembered fondly by older clasificadores at the *cantera*. Uncle Luca, whose house sits on occupied land directly opposite Usina 5, told me that Ponce 'gave us a hand, getting them to dump materials for us at night for a few years ... and coming almost every night to see how we were getting on'. 'He was our ally,' he explained, 'what a shame he didn't make it to mayor.' In 2002, on the tenth anniversary of Cacho's death, Ponce delivered a long eulogy in parliament and spoke of his time as director of public works. 'During the dictatorship, it had been common to confiscate clasificadores' carts and burn them, often in the *usina* – as if this would eradicate them!' he recounted. 'We changed that policy and had to promote another ... that of respecting and dignifying the work', he claimed, attributing this change to the legacy of the radical priest.

Ponce de León's last point is important: the GTC was not content with the legalisation of clasificador activity – it also sought to 'dignify' conditions of work, principally through proposals for collectivisation and formalisation. In order to act on waste and clasificadores, the new team had to construct both as objects of knowledge, combining surveys of clasificadores with studies of waste composition. In 1996, the group coordinated and published an Intendencia report in conjunction with the United Nations Development Programme (UNDP, in Spanish PNUD) entitled *Classification and Recycling of Solid Waste*, which included a detailed study of the sector and proposals for 'improving the socio-economic conditions of clasificadores', who were then estimated to number approximately 3,500 (IMM/PNUD 1996). Clasificadores emerged as a population in the Foucauldian sense, a 'mass of living and coexisting beings who present particular biological and pathological traits and who thus come under specific knowledge and technologies' (Foucault 2007: 367).

The content and language of the report was noticeably influenced by the agenda of the Rio '92 Earth Summit. It featured advisers from the Brazilian city of Curitiba, held up as a model of sustainable urban design and management, and a quote from Rio's Agenda 21 on the title page of the first chapter advocated an 'effective strategy to attack the problems of poverty, development, and environment simultaneously' (IMM/PNUD 1996: 1). It emphasised the need to involve clasificadores in 'a slow evolution towards participative and associative production that allows for their productive and social inclusion' (1996: vii) and called for 'a strong emphasis on activities of technical assistance, training, and accompaniment of clasificadores who voluntarily join the proposed experiments' (1996: viii). A first classification and recovery plant was envisaged for the clasificador-heavy Aparicio Saravia neighbourhood, to be followed by others in different parts of the city as the scheme expanded and more homes collaborated. In the plants, clasificadores would be 'accompanied and organised by NGOs', which would slowly hand over control of the plants to the workers once they were adequately trained (1996: vii). Informal economic activity, meanwhile, was to be discouraged and solid waste reconceptualised as 'rubbish that is not rubbish' (*basura que no es basura*) and the 'raw material of an industrial process' (1996: 210). In effect, the logic of recovering value in discards, long the operating principle of clasificadores, was to be appropriated by the state, and the operational function of containers was to be reprogrammed from hygienic enclosure to the enclosure of value.

The project's environmental impact assessment boasted of the positive impact changes it would have – not only on the environment but also on the clasificador: improvements in self-esteem, skills, personal development, and how they were perceived by others (1996: 209). Alongside 'more efficient recovery of recyclables from the waste-stream', the urban environment and the clasificador would also recover from the ills inflicted by informal work. Classifying in recycling plants rather than at home, the clasificador population would have less exposure to 'infectious and gastrointestinal diseases', engage in less strenuous physical labour, and enjoy a more hygienic environment at home and in the neighbourhood (1996: 210). These aims seem to hark back to governmental designs at the 'birth of biopolitics' in the nineteenth century, where the management of the population required 'a health policy capable of ... preventing epidemics, bringing down the rates of endemic diseases, of intervening in living conditions in order to alter them and impose standards upon them' (Foucault 2007: 367).

The report represented a shift in logic, from the repression of the sovereign, to the management of the clasificadores as a (vulnerable) population. Whereas for the dictatorship, rummagers and residues were a socio-material problem to be jointly eliminated, clasificadores and waste brought together under the policy banner of sustainable development could both be recovered into the formal economy. Waste-workers were no longer seen as infra-human but as a measurable population with a clear socioeconomic profile, concentrated in particular areas, and suffering poverty, deficient hygiene and high risks of disease. Capable of recovery and dignity if collectivised and brought into the formal sector, they were nonetheless considered incapable of autonomous production, at least initially, and thus in need of 'accompaniment' (*acompañamiento*) by appropriate NGOs.

Although the findings of the 1996 IMM/PNUD report were not implemented immediately, I dwell on them at length here because they laid the basis for the future Ley de Envases plants. A focus on 'accompaniment' emerged partly because of the influence of Catholic social activists in developing waste-picker policy, and this carried into the institutional design of recycling plants like Aries. 'We contracted organisations with a strong socio-educative background to coordinate the plants', noted one municipal official, 'because we took into account that although the workers had experience with classification, this had been gained in very precarious and informal conditions' (IM 2015: 109). Each recycling plant was managed by a different NGO: most of these had religious backgrounds and more experience managing vulnerable populations than waste infrastructure. The NGO given responsibility for managing Planta Aries – Cristo Para Todos (Christ for All or CPT) – was one such group.

Clasificadores at the plants were officially classified as 'protected workers', a move that had several implications. It entailed relatively flexible workplace discipline, at least during a transition period from informal to formal sector work. It meant the provision of socio-psychological support, so that several social workers formed part of the NGO team in Planta Aries. Finally, it meant that workers were given other forms of support – such as the possibility of finishing primary school or taking skills courses in areas like computing – aimed at making up a qualifications gap. One of the NGO plant coordinators, Richard, conceptualised his role in the plant as one of 'accompaniment and building citizenship'. In a presentation given to Aries workers in August 2014, María del Carmen inserted the associated benefits of their formal work into a long narrative of workers' struggles and victories in Uruguay, where the historic injustice of waste-pickers being left out of the labour-based citizenship of the Battlist state and its successors was finally being addressed. If Neilson and Rossiter (2008) write of the 'death of the citizen worker' in twenty-first-century Europe, in Uruguay it is more accurate to speak of her rebirth.

This particular citizenship project involved a retraining of the senses. As Gidwani (2015: 591) notes, 'livelihoods in the urban need economy implicate a "political history of the senses"', where 'what appear to be banal and dirty jobs … require a sensory resilience (visual, aural, olfactory and tactile) that is extraordinary'. 'That which bourgeois sense condemn as "filthy" or "revolting"', he concludes, 'is very often the normal order of things for the city's underclasses' (2015: 591). Classen et al., meanwhile, remark that 'evoking or manipulating odour values is a common and effective means of generating and maintaining hierarchies' (1994: 8). The problem for *acompañantes*, as framed by the Intendencia's head of social work, was that clasificadores had naturalised the foul smell of rubbish. In an inversion of Orwell's famous aphorism, the issue here was that 'the lower classes didn't smell', that is to say, they had apparently lost the capacity to distinguish, and be disgusted by, trash odours. Some clasificadores agreed, with Ruso from the *cantera* telling me that *gateadores* 'lose their sense of smell: all they smell is rubbish'. But when the head of social work celebrated one worker's complaint about foul smells within Planta Aries as an indication that his senses had been re-awakened, and of his inclusion in the sensory norms of the body politic, he retorted that the plant was simply smellier than the *cantera* because odours were concentrated in a restricted indoor space. It is worth noting here that, living and working around the landfill, both myself and my partner naturalised the smell of rubbish and learned to distinguish different smells, including burning plastics, organic waste, and the smell of the landfill itself, which seemed to thicken in the morning dew or with drifting fog and was not entirely unpleasant.

Some male workers flaunted their experience working in 'real' jobs in the formal sector, resented special treatment, and preferred plant work stripped of its citizenship component. For some, official classification as protected workers represented a challenge to their masculinity and autonomy. There appears to be little research into the threat that incorporation into the formal sector can pose to male workers' sense of masculinity in different circumstances, something that is perhaps a consequence of the (allegedly) greater focus on the recent feminisation of the global labour force (Mills 2003: 55). One exception is Phillipe Bourgois's (1995) classic ethnography of Puerto Rican crack dealers in Harlem who resist entry into service sector jobs because of low pay, demeaning conditions, and obligations of deference (often to female superiors) incompatible with their understandings of 'respect'.

In the case of my field site, and unlike studies that focus on vulnerability in relation to workplace injury (see Walter et al. 2004; Ye 2014), demonstrating vulnerability was not cause for losing a job in a recycling plant: it was effectively a pre-requisite for gaining one. From the outset, the provision of socio-psychological and educative services caused a split in the workforce between those who embraced and those who resisted them. This division was to a large degree gendered, with many men resenting what they interpreted as moves to infantilise them. 'They treat us like children,' complained Hojita, who joined several men in going out to smoke whenever socio-educative activities took place. When I asked CPT coordinator Richard whether such an attitude might have less to do with clasificadores per se, and more to do with generalised male working-class ideas of work, he agreed that 'workers prefer to be working' and that he might encounter the same attitude in a meat-packing factory, the Uruguayan proletarian job par excellence. In fact, Hojita and some others objected to being managed by an NGO altogether, arguing that:

> They [the Intendencia] contract someone who doesn't know anything to manage us. Neither us nor them: they outsource management. If I have a heart attack, I want a heart surgeon to operate on me, not a paediatrician. Who is better trained than clasificadores to know about rubbish? But they contract an NGO ... who have never opened a rubbish bag to eat or to get something out for the kids. I see no sense in it.

Hojita disparaged the professional qualifications of those involved in the social management of waste and waste-pickers, considering them an unnecessary bureaucratic layer providing care services that he neither needed nor desired. The NGO and the institutional bodies involved in implementing the law represented multiple levels of intermediation that, in effect, replaced the small-time neighbourhood intermediary the Ley de Envases sought to exclude by selling directly to industry. Not only did such institutional intermediaries necessarily charge fees that might otherwise have gone directly to clasificadores, they also made the management structure of the plants rather opaque, with workers unsure which institution could solve particular problems or disputes. 'Our problems [at the plant]', recounted one worker, Julia, 'are due to the number of institutions that accompany us: each has different responsibilities and this is difficult for us to understand' (IM 2015: 145).

Hojita and his colleagues were also unhappy with the newly established division of labour in the Intendencia, where responsibility for clasificadores had passed from the operational waste management to the department for social work. This meant that they found it harder to get hold of waste officials with relevant technical knowledge of the composition and provenance of waste trucks. Clasificadores had enjoyed logistical negotiation and the exchange of such knowledge at the landfill and COFECA. Landfill foreman Molina was almost in awe of *gateadores*, telling me that they were 'very intelligent, more intelligent than us'. 'I've put in ditches so that they can't get past', he explained, 'and they have responded with inventions that an engineer couldn't have come up with'. Part of the reason that non-literate forms of knowledge were appreciated by such municipal workers was because many came from backgrounds where manual skill was valued over and above academic qualification. 'Just because they didn't go to school doesn't make them less intelligent than us', Molina continued, 'my father didn't know how to read and write but he could still operate machinery [at the landfill]'.

At the recycling plants, however, certain forms of knowledge held by clasificadores – such as what *requeche* was fit for consumption – were devalued by institutional actors as practices thought to index an undignified 'culture of poverty' (Lewis 1966) that the plants sought to replace with a dignified and modern fixed wage. Hojita was worried about the loss of autonomy, income, and *requeche* that the move to the plant would entail. 'What we had at COFECA, we achieved with our own efforts,' he explained, referencing both workers' political lobbying to access valuable waste trucks, and their construction of the concrete classification platform at Usina 5. 'Now all we will get is the minimum wage.'

Yet compare Hojita's reaction to NGO management with that of Ana Clara, a middle-aged Afro-Uruguayan grandmother who combined work at COFECA with a small garment-making home business, often incorporating *requeche* cloth into designs that she sold at our local market in Flor de Maroñas. Ana Clara told me that had a COFECA representative been put in charge of the plant, she would have refused to work there. 'I want the NGO to stay,' she told me, 'because if the NGO goes then within a month everything will be completely filthy like it was at COFECA.' 'Unfortunately, we have to have a boss, a manager from the outside, because nobody followed the rules in the cooperative and that's how it ended in such a mess.' Ana Clara took full advantage of the different opportunities made available to her in the plant and obtained a primary school qualification.

Such differences of opinion were partly related to gender, as well as to the different backgrounds of those who ended up at the Planta Aries. One afternoon, I classified materials at Aries with Ana Clara on one side and Álvaro on the other. As we worked, Álvaro talked loudly with a male colleague about 'pussy' and the number of women that he would like 'to fuck', while Ana Clara listened uncomfortably. Ana Clara and Álvaro came from different worlds but had ended up together at the *cantera*. Both were structurally vulnerable, albeit in different ways. Ana Clara was a poor, female, Afro-Uruguayan rural migrant who had turned to the *cantera* after losing a job with a notary, Álvaro a *guacho* who had grown up around the landfill, never finishing school and becoming an intermittent user of various harmful substances such as *pasta base*. Support at the plant ostensibly catered to both trajectories and many more in between, but workers had to be open to some form of social, psychological, or educative intervention. While Ana Clara – skilled, entrepreneurial and upright – happily adjusted to life at the plant, Álvaro felt patronised and uncomfortable in a classroom environment, although he too was intelligent and hard-working. His priority was to earn enough money to care for his family but he also enjoyed the macho chat of the landfill, to which he soon returned. At the very least, he had gained a uniform out of his stint at the plant, and began to wear overalls emblazoned with the Ley de Envases's 'Your Packaging is Useful' (*Tu Envase Sirve*) logo at the *cantera*, mocking the government's failed attempts to incorporate him into a project of formalisation and social inclusion.

If schooling was an attempt to educate minds, then formalisation also entailed what we might call the 'vulnerabilisation' of bodies. Many landfill clasificadores thought of their bodies as having acquired resistance to microbes, and male clasificadores represented themselves as strong, muscular, and capable of carrying heavy loads on their broad shoulders. On entering the plants, not only was the consumption of *requeche* food frowned upon, bodies were also 'enclothed' in the garb of the modern proletariat: protective equipment like gloves, uniforms, boots, hats, back support belts, and protective masks. Heavy lifting was off limits, living labour replaced by the dead labour materialised in cranes and fork-lifts. Such changes posed a problem for male subjectivities built around the idea of strength and resilience. In fact, heavy lifting was so bound up with ascriptions of gender at the *cantera* that it recalls the work of Rita Astuti (1998) on the Vezo people of Madagascar. For the Vezo, Astuti argues, gender is not assigned at birth but performed in activity. Anatomically male Vezo could become female by adopting a 'female way of doing things' (1998: 42). The activity of lifting was one means of differentiating between genders, such that 'women carry heavy loads on their head, while men will always carry them on their shoulders' (1998: 43; see Figure 9). While I am not suggesting that female clasificadores could become male through lifting techniques or vice versa (although recall Gorda Bea's assertion that she was 'more macho than the men'), in both cases we can point to how people 'become gendered by way of acting and doing' (1998: 43–4).

In other words, the representation of their bodies as frail and vulnerable could be a challenging and emasculating experience for male clasificadores, even if the change in working conditions ultimately exposed their bodies to less risk. Take the example of Pedro San Román, one of four brothers who lived in Flor de Maroñas and worked at COFECA and then

Planta Aries. Only in his mid-thirties, Pedro already suffered terrible back pain from a decade of work heaving loads at Felipe Cardoso, and after a few months at Aries had to undergo a serious operation. One might think that he would have been happy to no longer have to constitute, at least in such a back-breaking way, the 'vital infrastructure' of recycling (Fredericks 2014). And yet I often found Pedro and others engaging in manual lifting and disregarding safety equipment when NGO staff weren't looking. The gender policing of 'hegemonic masculinity' (Connell 1995) certainly played a role in ensuring the continuity of such behaviour. For example, when Pedro uploaded a picture to Facebook of himself and his brother wearing new protective face-masks supplied by the NGO, I left a comment online asking if I could use the photo in an academic presentation. Another comment, from a fellow male worker, accused him of being a 'covered-up sissy' (*marica tapada*) and the two were soon engulfed in a digital slanging match, unknowingly rehearsing later (US) political divisions about mask wearing in times of COVID-19.

Workers were expected to take care of their bodies, but also of the plant itself, in a way that they had not done at Usina 5. This too caused problems along gender lines, with some men refusing to carry out tasks that they considered women's work, such as cleaning toilets. The more open, participative, decision-making spaces, monitored and facilitated by the NGO, also encouraged female workers to become more outspoken and assume positions of leadership. So, in a range of ways, male authority and subjec-

Figure 9 Gendered lifting practices at COFECA

Source: Author photo, 27 March 2014.

tivity was undermined at Aries. Most adapted, but many men who had been prominent leaders at COFECA returned to working at the *cantera*. It was not just men who left, however, several women did so too, and in circumstances that reveal the limits of this new form of Uruguayan paternalism and care at the plants.

Julia was from a large waste-picking family – her uncle had been a close confidant of Padre Cacho and both she and her sister Sylvia, who also worked at Aries, had been baptised by the priest, a moment captured in a photograph she proudly showed me. She became a trade union delegate at the plant ('because none of the men could be bothered', she told me), travelling to waste-picker conferences at home and abroad and featuring in the plant's promotional material. Yet when I returned to the plant a few years later, I was surprised to find that Julia had been fired. The NGO had been very tolerant of her lateness in the morning and her absenteeism, recognising its socio-psychological basis and providing appropriate support. But when CPT was replaced with another NGO, the latter proved less understanding and, after a string of unexplained absences, Julia's contract was terminated. Although she felt aggrieved and unsuccessfully asked MIDES to mediate on her behalf, other workers were less sympathetic, telling me that the extent of her absences made her position untenable.

At COFECA, Julia did not have access to socio-psychological support beyond that of her colleagues. But she could be absent for days or weeks at a time and still reassume her position at the cooperative. Such flexibility had been hard fought for by the cooperative, because the Intendencia was a lot more comfortable with a permanent workforce. In the formal workplace, however, such patterns were ultimately unsustainable, despite workers being 'protected' by a flexible approach to their conduct. If workers were excessively vulnerable and did not respond to socio-psychological support, then they were ultimately dismissed. Such a predicament compared negatively with the flexibility of waste commons like COFECA and the Usina 8 *cantera*, where a large pool of 'known faces' (*caras conocidas*) could work intermittently without ever facing dismissal, and differed from the kin-based labour of spaces like Natalia's yard. It also contrasted strongly with the street and its waste materials, to which clasificadores could also turn to in times of need. The creation of the Ley de Envases plants thus narrowed the criterion of vulnerability by which clasificadores could gain access to a waste livelihood. Further, as Don Kalb has noted, 'the state, bureaucracy and the rule of law turn a common good into a public good' (2017: 69) and indeed this is what occurred here, restricting the composite income streams previously enjoyed by informal clasificadores, as I go on to explore in the following section.

(Mis)Classification

Work at Aries consisted of processing waste delivered by rubbish trucks, which came from the new hermetically sealed, street and supermarket containers into which citizens were urged to deposit only recyclable waste. Workers divided into morning (6am–1pm) and afternoon (1pm–8pm) shifts that were spent picking materials from a conveyor belt and separating them into a range of plastics, papers, and metals. After classification, these were then baled by a small press that, along with the belt, forklift, and scales, made up the machinery of the plant. From the outset, Ley de Envases actors were over-optimistic about the volume and value of the recyclables that would enter the plants, and consequently the possible earnings of clasificadores, whose minimum wage was to be supplemented by the income from their sale. When the plants were launched the then head of Environmental Services at the Intendencia, fatefully declared to the press that formalised clasificadores would earn at least US$900 per month (*El Observador* 2014a). A DINAMA representative whom I interviewed told me that 'If a few orange peels end up in the containers and plants at the beginning, this will soon sort itself out.' Yet months after the launch of Aries, NGO coordinator Richard admitted that 70 per cent of the workers' time was spent separating and bagging discards. The skipping companies hired to transport such waste to Felipe Cardoso were surely one of the biggest initial winners from the implementation of the Ley de Envases in Montevideo.

Just as it was ironic that the Stericyclo employers we met in chapter 1 should hire former clasificadores to destroy value in the waste-stream, it was rather tragic that so much of Aries workers' labour was devoted to transferring rubbish from one bag to another. My colleagues found it positively absurd, and it made a mockery of the government's attempts to increase recycling productivity through a Taylorian division of labour (see Carenzo 2016). In the increased productivity expected to emerge from the enclosure of workers and the waste commons in these plants, we find a parallel with the justification for the enclosure of the English commons outlined earlier in this book. But whereas that enclosure, whatever its injustices, did lead to increased productivity, enclosure at the Planta Aries did not, as attested to by the steep drop in the income generated from the sale of materials when workers moved from COFECA. As the saying goes, 'garbage in, garbage out', and this was quite literally what occurred most of the time at the Planta Aries. But garbage was also coming in and out because of the garbage data that institutional actors had inputted when estimating how much of what arrived at the plants would be recyclable. As it turned out, a few bad apple cores were the least of their problems.

As well as miscalculating the value and the composition of the materials that would enter the plants, authorities misclassified the clasificadores who would work there. Different Aries workers articulated what they felt were a variety of mistaken perceptions of them. During one training session, Álvaro criticised NGO workers for thinking of clasificadores as 'Amazonian Indians', his particular shorthand for an ignorant, uncivilised people. It was another worker, however, Bolso, who elaborated the issue at most length during our interview, as we supped beer at the large Piedras Blancas Sunday flea market where he sold *requeche* with Hojita and Sergio:

> I wasn't born [at the dump], I finished school, I have a plumber's qualification; I'm not a nobody. They thought that we were extra-terrestrials, that we were cave-men, that we were from the ice-age. And it's not like that Patrick. Back there [in COFECA], I lived better, I had a better income than here. Since they saw us all dirty, 'Ah these guys are tremendous *píchis*, we'll give them 200 pesos and they'll be delighted.' That's what they imagined Patrick.

How fitting that Bolso, ever inviting us round the back of the plant to share a puff on a joint, should come up with such a trippy image of how the social apparatus viewed him and his colleagues: destitute alien Neanderthals. Yet as compelling as this figure might be, it is the latter part of his description which interests me here: that COFECA clasificadores, clothed in the dirt and smells of their workplace, were confused for the extreme poor. The level of misclassification varied according to state institution, so let us begin with an example from the organisation that had least contact with clasificadores generally: INEFOP.[2] This para-state organisation was contracted to deliver a series of 'transversal skills' training sessions to prepare workers for their move into the formal sector.

During one such session, Gordo Callao asked about parking facilities for his horse and cart, and was told by the workshop facilitator that these wouldn't be appropriate at the plant. 'What about the two cars I have in the garage?', he insisted. Parking needs had been 'estimated based on the type of people entering the plant', she responded, and clasificadores weren't expected to own cars. 'If we were seen building recycling plants for people with cars', the DINAMA official explained when I raised the episode in an interview, 'then we would be questioned by international organisations who would ask what this had to do with the eradication of poverty.' Clasificadores were thus damned if they did and damned if they didn't: horses and carts challenged norms of 'infrastructural modernity', while car ownership effectively challenged the criteria of economic vulnerability by which state actors could justify investment in the plants to international partners. While the Uruguayan welfare state was built in the early twentieth century around the figure of the (male) industrial worker (Pendle 1952), governmentality at the beginning of the twenty-first, in line with an international development focus on the extreme poor, now targeted the inclusion of the waste-picker.

Peattie's warning that '"the informal sector" is by no means equivalent to "the poor"' and that 'there is plenty of evidence that incomes among small-scale entrepreneurs cover a great range from extreme poverty to well over average' (1987: 857) remains relevant in Uruguay today. Workers like Hojita, Bolso, and Sergio often doubled their wages working as market-stall holders who sold *requeche*. Gordo Callao's cousin Sapo, meanwhile, had amassed capital as a burglar before he was 'born again' as an evangelical Christian. He also owned a car, as did his nephews Cholo and Porteño. In fact, Porteño possessed a large vintage Chevrolet truck, a beautifully restored model popular in Uruguay among fruit and vegetable sellers for whom he worked on market day. On other occasions, Porteño would borrow the four-wheel drive belonging to his father, a long-time glass bottle wholesaler who had turned his hand to importing goods from Brazil. 'Being a clasificador', Porteño told me, 'doesn't mean that you are living in the mud or eating out of a can, that's not dignified for anyone.' Perhaps to prove his point, he arrived at the plant's press-heavy inauguration in the four-wheel drive.

High earners in the Intendencia were suspicious of those who had made money as clasificadores in the informal sector, with the informal coming to be associated with the illegal or illegitimate. One told me that he reckoned Porteño to be a '*delincuente*' who had become rich by appropriating and selling glass bottles set aside by COFECA. Porteño had in fact openly classified and sold these bottles, and this had been permitted by the cooperative because of their low market value. But he made most of his money from removals (*fletes*) and the small shop that his wife ran out of their home. Porteño's cousin Enrique, who earned a lot of money labouring at the *cantera*, occasionally dropped workers off at Aries in his shiny new Volkswagen and was presumed to owe his wealth to the labour of peons working for him at the landfill, whereas in fact he mostly worked alone.

According to Bolso, state and non-state actors presumed that those entering the plants were so poor as to be appreciative of a hand-out. Yet given not only the heterogeneity of the clasificadores, but also the financial pressures on the poorest workers, low pay soon became an issue. Workers received a national minimum wage, which, after taxation and social security contributions, amounted to only US$340 per month, a sum that was widely considered low for a country with high food and transport costs. This was meant to be supplemented by additional income generated from the sale of recyclable materials directly to factories. For the first few months, as workers amassed enough materials to be sold and waited for the legal apparatus that would enable those sales to be finalised, the Intendencia agreed to create breathing space by giving them an additional stipend of US$150 per month, raising the monthly income to just under US$500. Yet this still amounted to a small sum if one remembers how much money could be made in the *cantera*, where many of the male clasificadores at Aries could enter and work freely.

In discussions with workers and in public statements, authorities emphasised how the work was made more dignified through the provision of 'a roof', improved health and safety conditions, and social security at the Ley de Envases plants. Yet as Bolso expressed when speaking beside Mayor Ana Olivera at the ceremonial hand-over of INEFOP training certificates: 'Work-wise it's great that we're under a roof, but just because we're under a roof doesn't mean that we should earn less.' When the sales of materials finally started to go through, these turned out to be worth less to the workers than expected, because although better prices per kilo were assured, the recyclable fraction of waste was less than foreseen and social security contributions of almost 50 per cent were deducted.[3] Income from the sale of materials averaged a mere US$100 per month per worker, compared to the US$100 workers earned *weekly* in COFECA.

Given the institutional focus on recognising workers' role as environmental agents – recognition materialised through organised trips to schools and businesses, media interviews, and indeed the construction of the plants themselves – we can suggest that the state also designed a 'waste wage' into the Ley de Envases. By this I do not mean a bonus received for unpleasant and potentially hazardous work, something that municipal waste-workers received but clasificadores did not. Rather, I mean something more akin to what Patrick Vitale (2011) calls the 'war wage' designed for American military and defence workers during the Second World War. 'The war wage', writes Vitale (2011: 785), 'was not a pecuniary wage, rather the state and industry offered a sense of sacrifice, contribution, and national belonging to workers and civilians who faced rationing, wage freezes, extended work hours, and emotional duress.' The 'waste wage' would thus also mean a non-pecuniary element of a composite wage (O'Hare 2013), with workers encouraged to put up with low salaries in return for their celebration as heroes who could take pride in a form of sacrificial labour that benefited not just the nation but also the global environment. As such, the national citizenship embodied in social security provision was complemented by pretensions to global environmental citizenship (see Stamatopoulou-Robbins 2014). Without downplaying the importance of such recognition for workers, they still had to put food on the table. 'We are clasificadores, environmental warriors [*guerreros del medioambiente*]', Hojita emphasised, 'and they are only going to pay us the minimum wage?' Hojita was thus proud of the important ecological role that informal and newly formalised clasificadores played, but for him this ideological 'waste wage' did not obviate the need for a dignified salary.

Financial pressures were aggravated by the explicit prohibition on taking home *requeche* from the conveyor belt, a bounty that was in any case much reduced because the plants were restricted to receiving only household waste. According to a representative from the CIU whom I interviewed, workers were not permitted to take anything from the plants because they might sell these on individually and this would mean 'fomenting the informal sector'. Since workers' wages came from voluntary contributions made by businesses that released non-returnable packaging into circulation, plant workers' labour should be spent recovering such packaging, not objects to be sold individually at the flea market. MIDES established a rule that materials should either be classified as recyclable and baled, or else placed in skips to be taken to Felipe Cardoso. There was room for *material* and *basura* then, but the key third clasificador category of *requeche* was excluded. Workers had feared such a development, with a worried Sergio telling me that the *requeche* taken from COFECA would 'not exist at the plant … if you find a watch, you'll have to hand it in; if you find money, you'll have to hand it in'. While I didn't miss the heavy lifting of the *cantera*, I certainly missed the variability, surprise and sense of getting something for nothing that were materialised in *requeche*. And when I asked Aries worker Joana what she missed about the *cantera* and cooperative, she responded:

> What do I miss? That in the *cantera*, for example, one day you didn't have anything to eat and from a truck you got chicken, meat ... bags of pastries, fruit. There was a truck of burgers from MacDonald's. You didn't have anything to eat and you took them home, stuck them in the oven and could eat burgers. These are things that you miss.

The prohibition on the recovery of *requeche* recalls a past situation documented by Peter Linebaugh. In *The London Hanged* (2003 [1991]), Linebaugh details how seventeenth-century London dock workers took home 'chips' as part of the remuneration for their labour:

> What were chips? What were they worth? Broadly speaking, they consisted of wood scraps and waste created during the work of hewing, chopping and sawing ship timbers. The term refers not to the wood itself but to the right of the worker to appropriate a certain amount of it – a prescriptive right since 1634. (2003 [1991]: 378)

Chips were, in other words, what clasificadores called *requeche*: the material leftovers of a particular process of productive labour. *Requeche* too, then, might be described not only as things – but as customary rights to them. And just as the recovery of *requeche* was often worth more to COFECA workers than the income they received from the sale of stock recyclables, so too 'the chief remuneration of yard workers was not their monetary wage' (2003 [1991]: 378), but chips, 'a perquisite providing between a third and a half of weekly earnings' (2003 [1991]: 379). Just as clasificadores used *requeche* in the construction and furnishing of their homes and to replace household expenditure, for 'those having a right to this prescriptive custom, chips were an essential part of their ecology – in housing, in energy, in cooking, in furnishings' (2003 [1991]: 379). Chips formed part of a wider moral economy of material surpluses – 'sweepings', 'overweight', 'gifts', 'the flows' – to which London workers had established customary rights and which had become 'a known and accepted part of the class relationship' (2003 [1991]: 406). As Linebaugh writes, 'customary appropriations appear as inefficiency or waste to the technologist' (2003 [1991]: 430) and they did so to the actors involved in the implementation of the Ley de Envases: a way that valuable labour time would be lost recovering materials for individual gain and valuable materials lost to the 'informal sector'.

The CIU's attempts to discourage informal sector activities are hardly surprising given that the trade body raises voluntary EPR (extended producer responsibility) funds from large businesses that all pay their taxes, at least in theory. When I interviewed then President of the Chamber of Uruguayan Waste Managers (CEGRU), he also complained about being undercut by clasificadores who collected waste without paying taxes and might even skip the cost of dumping in Felipe Cardoso. Yet is this crack-down on informal activity also representative of Uruguayan government policy and, as such, at odds with moves to celebrate small and social entrepreneurship in the European metropolis from where Uruguay draws its ideas of infrastructural modernity? In the UK, for example, much has been made of celebrating entrepreneurship, voluntarism, and small business in recent years. Yet the British government hardly celebrates the informal sector as such, and continues to push to regularise it, so as to improve revenues and prevent exploitation and modern slavery. In the recycling sector, the informal scrap metal trading was prejudiced by legislation like the Scrap Metal Dealers Act (2013), while in the waste management business Steve Davies (2007: 16) has shown how, in spite of Thatcherite rhetoric stressing the importance of small businesses and the entrepreneurial spirit, reforms such as the Local Government Act 1988 actually transformed a sector dominated by small operators into one run by very few multinationals. In Uruguay, the situation was somewhat reversed: the centre-left Frente Amplio rarely talked up entrepreneurialism or small business but its formalisation of the waste transport sector, as well as disenfranchising informal clasificadores, multiplied the number of registered '*unipersonales*' or sole traders.

Back at the Planta Aries, punishment for the recovery of *requeche* at Planta Aries did not materialise. Like the first attempts to restrict the removal of chips in England, plant coordinator Richard made sure that the prohibition on the removal of *requeche* was a 'dead letter from the start' (Linebaugh 2003 [1991]: 378). Charismatic, from a large working-class family in Flor de Maroñas, and with a clasificador brother himself, Richard simply told MIDES that it was more dignified for workers be allowed to remove things from the conveyor belt than have them rummaging in the bins, which would surely happen if valuable items were placed in discard skips. So, workers were allowed to jostle to load materials onto the belt, since this afforded them first choice on any *requeche*, which thus continued to circulate on weekly stalls at Piedras Blancas. Workers like Bolso even managed two separate categories of *requeche* from their place at the belt: one for food scraps that he gave to a neighbourhood dog-owner in exchange for marijuana; another for objects that could be sold at the market (see Figure 10). The continued circulation of *requeche* was something of an open secret: on one occasion, a worker wandered cheerfully past a meeting of MIDES, CIU and DINAMA representatives with a roll of cables, letting Richard know that he was taking them 'in case he thought he was stealing'. 'What a moment to say that!' smiled the plant coordinator, embarrassed.

Figure 10 A pile of *requeche* recovered from the conveyor belt at Aries
Source: Author photo, 27 May 2014.

The presence of informal sector economic activity went beyond the recovery of *requeche*. Angered by the limitations of formal sector sale, where a buyer could not be found for the glass or bottle caps stockpiled at the plant, or for the *nailon negro* (black 'nylon') that became wet and rotten as it sat outside, workers called upon old contacts. Despite objections from the Intendencia, they telephoned an intermediary and sold him the glass bottles, which he paid for in a wad of cash that workers distributed among themselves. If Richard permitted the exit of materials into the informal sector, he also admitted the entry of trucks containing materials that did not come from approved sources. As we have seen, trucks were only meant to contain domestic recyclables but workers demanded the right to receive valuable commercial waste, even emblazoning the request on the mural that they created with my partner Mary on an exterior wall: 'The Intendencia (or whom it may concern) should allow big businesses and shops to give us what they don't need, and what we can use [*nos sirve*]'.

In planning and constructing the mural, my partner incorporated the perspectives of workers so as to nuance the purely celebratory vision of the move to the plant that had been envisaged by the NGO. Although this was her first community art project, she drew on her previous experience of using discarded and recovered objects in her practice. The mural featured a mixture of plastic bottle caps, aluminium cans, and painted figures inspired by photographs taken at COFECA and at the Planta Aries, as well as a series of small panels designed by individual workers. These were joined by recovered tyres that been transformed into outdoor furniture to be used by workers while on their breaks, and a path traced out with glass bottles set into concrete. Within the painted representation of the large *bolsones* that clasificadores filled with recyclables, the workers painted a series of slogans, highlighting the dignified nature of their work, their corresponding right to a dignified income, and the environmental service that their labour constituted.

This demand to receive commercial waste highlights a point largely overlooked not just by Ley de Envases authorities but also arguably in the wider discard studies literature: it is not only that the largest volumes and greatest hazards of waste are to be found in industrial-commercial rather than household streams (Harvey 2016: 7) – such streams can also contain the most value as well. In fact, the formal sector recycling of household waste at a global level is not particularly profitable and invariably relies on state subsidies (MacBride 2011). In line with a global focus on educating and 'responsibilising' individual citizens to ensure the efficiency of recycling (Hird 2017: 190; MacBride 2011: 4), institutional actors blamed the poor quality of the materials that arrived at the plants on Montevideans' lack of recycling culture. But clasificadores knew that even well sorted household waste could not rival the scale, homogeneity, and value of the industrial-commercial waste that they had regularly received at COFECA. Such waste was not included in the Ley de Envases and did not generally arrive at the plant, but when it did, Richard accepted it. 'If a truck comes with useful material for the plant, it'll enter,' he explained. 'I know that the plants were built for the Ley de Envases but today the law isn't covering the workers' needs ... we can't bury a truck of paper just because it's not covered by the law,' he added. Institutional attempts to enclose all materials within the formal sector as part of the Container Law thus quickly came undone. The leakiness of waste, a point recognised by various waste scholars, (e.g. Harvey 2016: 7; Hird 2012) thus extends to its ability to seep out of the formal and back into the informal economy as valuable *material* and *requeche*.

It has been argued that the informal sector globally is often made up of economic units that are subordinate to formal capital; that its workers serve to 'reduce input and labour costs of large capitalist firms' (Chen 2012: 488); and that informal economic activity can thus be seen as tantamount to a 'regime to cheapen the cost of labour in order to raise the profit of capital' (Breman 2013b: 1). In Uruguay, Sarachu and Texeira (2013: 4) argue that informal sector *clasificación* is a 'productive complex' where 'the enormous profit margins of recycling firms are based on the hidden exploitation of the work of the clasificador'. What proves interesting at Planta Aries is the way that informal economic activity continues to subsidise the formal sector, even within a formalisation scheme that explicitly sought to displace it, while formal sector businesses involved in the transformation of recovered materials continued to profit from those who got their hands dirty lower down the supply chain.

A final case of potential misclassification surfaced during a plant visit from the large paper buyer IPUSA, based in the outskirts of Montevideo and involved in the transformation of used paper into such Uruguayan household brand names of toilet paper, sanitary towels, and nappies as Hygienol, LadySoft, and BabySec. The company represented 'the factory' (*la fábrica*) that the plants were trying to sell to directly, thereby bypassing the chain of intermediaries to whom COFECA had previously sold white paper. IPUSA, in turn, was attracted to the plants because they realised that, although the formalisation of waste-picking only represented a small part of the market initially, consolidation of the model would substantially shake up the recycling trade.[4]

Yet the quality of paper that the company had begun to receive from the plants was sub-standard, and representatives were sent to Aries to let workers know where they had been going wrong. With the visit, two links of a chain normally separated by several levels of intermediation were meeting for the first time. The visitors were technicians from the operational team that dealt with the mixing of raw materials to produce the paste from which new products were elaborated. Instead of the clasificador paper category of *blanco* (white), IPUSA dealt with the categories of White 1, 2 and 3, which differed accorded to the amount of print the page contained (none, little and lots). They classified coloured paper into 'special mixed', 'mixed', 'magazine', and 'punched'. In the process of making new white-paper-based products, the whitest and most unadulterated materials were of the highest value in this fine-grained system of classification.

Workers who for years had prided themselves on their classificatory skill and whose very occupation had come to be known as 'classifier' were told that they had been doing it all wrong and had to retrain their perception of materials. 'You don't have the knowledge', explained the factory representative, 'you are putting a lot of things into the bales which shouldn't be there [but] the idea is that you get trained up ... and improve.' The new way of conceptualising materials was met with a mixture of scorn and disbelief by the assembled workers. 'It would be very difficult for us to sell you those categories ... we just sell white and mixed paper,' responded Ratón. 'I don't think it's going to work,' Michael later told me hesitantly, 'we're used to a different type of classification ... one likes to come in, classify roughly, make a little money and go home.' Still, he said that workers had 'started trying to separate out some papers that they [IPUSA] said didn't go but which for us had been white all our lives'. Paraphrasing Demos (2013), clasificadores were asked to 'look again in a different way' (2013: 114), and reconfigure not just their olfactory but also their visual encounter with discards.

Most of us would side with clasificadores in holding up a piece of white paper and affirming that it was, indeed, white paper. Such a predicament is reminiscent of George Lakoff's assertion that 'we have a folk theory of categorization itself, which holds that things come in well-defined kinds, that the kinds are characterized by shared properties, and that there is one right taxonomy of kinds' (1987: 121). The white paper in this instance is precisely *not* what Bowker and Leigh Star (2000) call 'boundary objects': 'objects that inhabit several communities of practices ... that are able to travel across borders and maintain some sort of constant identity' (2000: 16). No such object as a generic white paper could make its way up to industry: parts of the clasificadores' *blanco* would be subclassified into different categories, other parts would be rejected. The meeting between industry and clasificiadores entailed a clash of different classificatory systems that in fact were usually invisible to each other because of the actions of intermediaries who translated between different social worlds and standards.

Paradoxically, one thing that both groups of classifiers seemed to agree on was the low value of household waste: the very material whose exchange brought them together. 'For us, household waste is organic waste because it contains food, used toilet paper and *yerba* tea, and we vulgarly call it "dustbin" [*tacho*]', explained the IPUSA representative. He added that 'Uruguayans, regardless of social class, lack a classificatory culture [*cultura de clasificar*]'. Because of this, he continued, the waste arriving at the recycling plants was not neatly separated, unadulterated, raw material but contained many elements of this 'dustbin waste'. The possibility of finding paper that had not been dirtied with *yerba* tea leaves or cardboard on which oil had not been spilt was thus slim. 'You're talking as if the materials all came like that,' one worker observed, gesturing to a book of sample materials that IPUSA had brought to the plant, 'but everything comes filthy [*mugriento*].'

In leaving the samples, the representative told the workers that this was 'the classification that we want you to achieve, this is happiness … you may not reach happiness but we want you to get as close as you can'. Like alchemists, workers were expected to extract from the household waste-stream a utopia of industrial-quality classification. Instead, they spent most of their time repacking rubbish. The advertising campaign for the Ley de Envases had probably not helped, featuring as it did packaging icons like fast food burger boxes and drinks cartons that could not, in fact, be recycled locally. But the principal reason for the failure to get closer to classificatory nirvana was the prior legislative classification of waste into household, commercial, and industrial streams, from which recycling plants could receive only the first: 'dustbin'.

Containment

The material presented thus far suggests that the Ley de Envases plants both continue and reconfigure prior dynamics of commoning and enclosure. They maintain but reorder the link between vulnerability and waste but also represent an attempt at hygienic enclosure, where workers are shielded from the risks of places like COFECA, are materially and symbolically 'enclothed' in the protective garb of the formal worker, and receive a restricted waste-stream that ideally limited potential exposure to hazardous material. In this final section, I develop two further examples of enclosure. First, I outline the attempt to contain what we might call workers' 'excessive masculinity', where CPT was charged with managing and containing workers' behavioural excesses. Second, I note how containers were re-engineered to protect resources from criminalised appropriation rather than protect the public from contaminating rubbish. Such a rechannelling of waste is in turn leading to a possible 'tragedy of the commons' at the landfill, to which many disenfranchised kerbside recyclers were forced to turn.

As I have already noted, perceptions of the move to Aries were heavily gendered. Many women matched a general enthusiasm for the plant with support for its management by an NGO. Initially suspicious about NGOs that 'took clasificadores' money', La Negra changed her mind after becoming tired of the superior airs adopted by some male cooperativists and their unwillingness to open the books at COFECA.[5] Both La Negra and Ana Clara spoke about how reluctant they and other workers were to take orders from COFECA colleagues, with a little more respect afforded to 'someone from the outside' (*alguien de afuera*). Just as women had benefited financially from the change from individual to cooperative work at Usina 5, they hoped for better conditions and changed leadership dynamics at the plant. As it transpired, most representatives continued to be men, at least initially, but women became more outspoken in political discussions. Julia's sister Sylvia, for example, at one point came forward nervously and with trembling hands offered a list of group demands she had written down to be included in the plant mural and in a letter to the mayor, and was surprised when these were adopted for both purposes.

Some of those in structurally similar conditions, such as retirement-age men Sergio and El Abuelo, had radically different opinions about the move. While El Abuelo emphasised how lucky he was to be given a job with social security benefits at his age and asked how colleagues could possibly miss working in the dirt and exposed to the elements, Sergio said that he would miss 'everything' about COFECA and worried about losing his new job immediately. COFECA had a special meaning for Sergio because, in contravention of the Intendencia's rules, he had actually slept at Usina 5 with his cats and alongside a rotating gang of male clasificador colleagues who, temporarily dumped by their wives, hesitantly made their way along Felipe Cardoso to bunk up alongside him. 'Being bossed around [*mandado*] and taking orders when I've always worked as my own boss will be difficult', Sergio worried. 'If they come and shout "you can't smoke, you can't go to the toilet three times, you can't stop work" at me, how long can I last in this business? Two days? And then what do I do?' In the end, Sergio managed to stay on, despite earning himself a week's suspension for slapping a (male) NGO worker in a dispute over the consumption of alcohol in afternoon breaks. But years of self or cooperative employment led to a disdain for those who worked directly for others, whom some referred to as 'dogs'. Indeed, one clasificador complaint was that they were to obey orders given not directly by an employer or a key stakeholder in the Ley de Envases (such as the CIU, which paid their wages) but by NGO workers, intermediaries who themselves worked for someone else.

In one training session, Álvaro expressed these feelings through a visual medium. Workers were asked to create a collage that depicted their sentiments towards the transition to formal work and in his group, Álvaro cut out a dog's head from a magazine and, with two arrows, signalled that the animal represented two plant coordinators. A speech bubble had the supervisors chastising workers with the orders, 'Callao, don't shout!' and 'Álvaro, don't say that!' At the Planta Aries, the dog had apparently become the master. Other images on his collage had clearly been cut out from the textbook of Uruguayan masculinity: a football emblazoned with the Uruguayan flag and a scantily clad woman saying: 'I've not been given a uniform, should I work like this?' In the following weeks, more sexist behaviour occurred in the plant where, unlike at COFECA, it registered as unacceptable. First, explicit sexual language was used between men, in front of and towards women, in the form of outbursts, lewd stories, and unwanted sexual advances. Then, notes were left in the lockers with drawings of penises and sexual insults; a banana was left suggestively in one woman's locker; and the women's changing room door was opened slightly by a man who exposed his genitals to the woman inside. A masculine 'aesthetics of vulgarity' (Mbembe 1992) resurfaced at the plant.

These incidents of sexual harassment occurred simultaneously with acts of aggression carried out against the plant's security guard. Although, or perhaps because, they generally came from a similar social milieu, clasificadores tended to hate security guards almost as much as they hated the police. One reason for this, alongside the repression that they suffered in their neighbourhoods and at the *cantera*, was that some of the male COFECA workers had suffered from another form of enclosure – incarceration – and were thus accustomed to a prisoner–guard style antagonism that they carried into the plant. Ricardo, the plant's first security guard, was seen as something of an easy target, and in one incident his bag was urinated into. On another occasion, Callao was asked to vacate the security guards' chair in which he had sat down, and he replied that if he wasn't allowed to sit there then the next time the security guard tried to sit with the workers, he would smash the chair over his head.

Callao had been brought up in and around the *cantera* and was the younger brother of Enrique, the landfill's highest earning clasificador. At the plant, he had received a warning for boisterously singing and dancing his way through shifts. He had also taken a dislike to a young NGO worker who came from the neighbourhood bordering the *cantera* and had threatened him with physical violence if their paths crossed on the outside. This supervisor had failed to establish his authority with the workers, and some openly insulted him, with Callao telling one colleague not to give him sugar for his coffee because he was a 'cock-sucker who earned at least double what they did'. Callao was suspended after the chair incident and his friends in the plant organised a walk-out in solidarity with him. Yet here too, gender dynamics were at play. Some women claimed that some of the 'heavier' male figures in the plant threatened that their female kinsfolk on the outside would assault them if they refused to join the walk-out. Some female workers had sympathy with Callao's adversaries. 'Imagine you are the boss and you're being laughed at in the face with that laugh Callao has, that mocking laugh, and you realise you are being ridiculed,' said one.

Yet overall, I found the workers in Aries visibly subdued following their move from COFECA, at least initially, and when I enquired as to why, some said that this was because their boisterous behaviour at work, even their liberty, had been curtailed. 'They've taken a lot of freedom from people,' Bolso told me, 'the other day they even said that we couldn't sing and dance. Because we sing and dance at the *pista*, you know? We have a laugh … when I go to get changed, I shout at the top of my voice to everyone!'[6] Most of the shouting and singing was neither obscene nor misogynistic but it was rowdy. Many workers would sing along to *cumbia* music on the radio; and the popular Matute would often call out wildly to one of the older female workers who would respond by leading the dancing at the conveyor belt. Clasificadores were also used to establishing joking and teasing relationships with one another and with intermediaries. So when the visiting IPUSA representative said that egg boxes weren't good for anything (i.e. couldn't be recycled), Ratón quipped that they were only good for 'breaking eggs' (*rompiendo los huevos*), a play on words since the expression also means to 'break someone's balls'. The joke was met with laughs from his colleagues but the silence of the visitors who continued to dryly enumerate a list of potential paper adulterants.

The discouragement of noise and 'obscene' gestures in Uruguayan industry has a long history. Barrán writes that at the turn of the twentieth century in Montevideo, 'where a modern factory was established, the norm was work in silence and prohibition of all "racket"' (2014: 397). The 'new sensibility' of the industrial bourgeoisie considered silence – at least in the lower classes – as good taste, while 'tuneless and shrill shouting and obscene gestures revealing bodily needs should be hidden' (2014: 397). 'Dirtiness, uninhibited gestures, guffaws and unruly shouting had co-existed and been highly regarded during the "barbarous" period,' he writes, while 'neatness, bodily discipline and whispering or silence were prized during the period of "civilisation"' (2014: 398). 'Good taste', argues Barrán, 'correlated suspiciously with the interests of the industrialists [*patronato*]' (2014: 397).

In *Rabelais and his World*, Bakhtin (1984 [1965]) famously argues that the grotesque and the obscene were deeply embedded in plebeian life, and he counterpoises the closed impenetrability of the bourgeois modern body with the porous one of 'genitals, bellies, defecation, urine, disease, noses, mouths and dismembered parts' (1984 [1965]: 319) that predominated in the humour of the European folk tradition. 'Whenever men laugh and curse, particularly in a familiar environment,' he notes, 'their speech is filled with bodily images' (1984 [1965]: 319). To the extent that the *cantera* was a familiar environment to clasificadores this can also be said to be true here, but male clasificadores also policed the tone and language they used in the masculine work environment and in the familial domestic sphere. Álvaro, for example, told me that he was a different man at home, and when I met him with his wife and kids, my boisterous co-worker did indeed cut the figure of a meek, respectful, family man. Both the state and male clasificadores thus sought to keep dirt in the workplace and out of the domestic sphere, only for the former this meant dirty labour and the latter dirty language.

Mbembe (1992) adapts Bakhtin's articulation of politics and vulgar parody, arguing that in their use of body-based humour, subjects of the 'post-colony' do not necessarily challenge state power, since 'for the most part, people who laugh are only reading the signs left like rubbish in the wake of the *commandement*' (1992: 10).[7] For my part, I seek to avoid reproducing the dualism inherent to both Bakhtin's modern/ popular body imagery and Barrán's account of the move from barbarism to civilisation in Uruguay. But unlike in Mbembe's (1992) case, the locus of vulgarity and humour in the recycling plants still lay largely within the sphere of popular culture rather than state power. The (masculine) clasificador body danced, laughed, sang, swore, drank, ate *requeche*, and pissed, in one instance in the poor security guard's bag. This act mirrored one that had occurred a few years earlier at the *cantera*, when young *gateadores* had defecated and rubbed their faeces over a police cabin in protest at attempted enclosure, and echoed that which took place during the dictatorship, when arrested clasificadores had filled their boots with stinking rubbish and emptied them in police cells.

Although Hojita compared the plant to a barracks, and one can find some similarity in the historical descriptions of Barrán (2014), it is important not to exaggerate the disciplinary measures and climate that operated at Aries. As we have seen, permissiveness and patience were central to the NGO's approach to these 'vulnerable workers'. Institutional actors did not try to engineer permanently silent or solemn bodies. Instead, spaces for loud, boisterous, and joyous behaviour were to be contained in particular space-times that were outside of working hours and in keeping with this twenty-first-century attempt at the social inclusion of the vulnerable worker-citizen. MIDES organised two such spaces for Ley de Envases workers during the course of my fieldwork. The first of these was a 'social tourism' trip to a river and park, with the group entertained by loud music, games, and a buffet lunch. The second trip was to a national conference for clasificadores working at Ley de Envases plants around the country, and featured a live band at lunch, which played covers of popular 'tropical' songs and soon had riotous clasificadores (and then *técnicos*) on their feet dancing. This was the highlight of the excursions for many clasificadores, and a MIDES official told me that they had made sure to include live music (and generous servings of food) following feedback from clasificadores at previous events.

Still, there were moments during such events where popular humour pushed the limits of institutional acceptability. On the social tourism trip, for example, a game was organised whereby the bus was divided down the middle and each side had to try and get the maximum number of balloons onto the other side before the music stopped. Yet Sergio decided to create his own fun by bouncing a balloon off the back of the head of a municipal social worker for large sections of the journey. On the same trip, Ratón ordered large amounts of meat to be grilled at an all-you-can-eat buffet, purposefully creating leftovers that he then bagged up to take home. 'It's for my dog', he joked, to the laughter of his colleagues on the bus home, before letting out a loud woof. Even outside the plant and in peaceful country surroundings, Ratón had managed to provoke an episode involving the contested consumption of *requeche*.

Effectively, the majority of workers at the plant 'voted with their feet' (Guha-Khasnobis et al. 2006) on the process of formalisation by staying at the plant, while dominant men did so by returning to the *cantera*. Work at the *cantera* was also under threat by another more direct form of enclosure, however. Disenfranchised kerbside clasificadores who remained outside of formalisation schemes began to appear there, bringing with them greater competition for materials and undermining the fragile détente that existed between municipal workers and my *gateador* friends. According to the UCRUS president, numbers at the *cantera* had rose from 50 during my fieldwork period to over 200 in 2017 because of the difficulties waste-pickers had in accessing waste in containers and entering certain areas of the city (López Reilly 2017). This increase in numbers in turn brought increased media attention and complaints from municipal employees, leading the Intendencia to announce that clasificadores would be excluded from the site (López Reilly 2017). The attempt to lessen precarity for small number of workers (128) in the plants has increased it, I would suggest, for the much larger number (3,000–9,000) for whom there is no room in the 'craft-in-the-making' (Carenzo 2016) of formalised waste classification. It also increased the chances of the tragedy that the enclosure of an overpopulated *cantera* commons would imply for scores of families.

To understand such outcomes, we might turn to Kasmir and Carbonella's (2008) idea of 'dispossession as the production of difference'. The labour scholars argue that political dispossession is 'frequently compounded by structural violence connected to the recategorization and reclassification of working classes' (2008: 13). One example they give is the recategorisation of the London riverine poor in the eighteenth century, documented in the work of Peter Linebaugh (2003 [1991]). As we have seen, the previously common practice of workers receiving wood from the shipyards as a form of payment was outlawed, creating in the process a split between the waged and unwaged, so that 'the literal policing of the division between waged labourers and the wage-less poor effectively separated the struggles of workers within the labour process from those outside it' (Kasmir and Carbonella 2008: 14–15). Yet it is important to note that this dialectic of dispossession through the redrawing of the boundaries of capitalism and waged labour is neither a historical artefact nor relegated to the margins but is rather integral to contemporary capitalist processes and the reconfiguration of class dynamics.

In Montevideo, dispossession-by-differentiation was justified by a fracturing of the clasificador trade in an example of what Samson (2015b) calls 'hermeneutical injustice'. Samson uses this term to describe the way that municipal officials at the South African Marie-Louise landfill referred to those who recover waste there as 'scavengers' instead of their chosen term, 'reclaimers'. Such naming practices, Samson argues, 'enable municipal governments and private companies to dismiss them [sic] as nuisances who need to be eradicated' (2015b: 825). When I interviewed an intermittent director of Felipe Cardoso, he engaged in a similar act of 'hermeneutical injustice' by insisting on referring to landfill waste-pickers not as clasificadores, as they wished to be called, but as *hurgadores* (rummagers):

> We have a problem with the rummagers [*hurgadores*]. A rummager is he who rummages. Let's call a spade a spade. Rummagers don't like being called rummagers but whoever is in the landfill is rummaging to see what he finds. The word rummage means to search. They say 'we are clasificadores'. Well, maybe those who work at the plants who classify and work in a certain way, we could call them clasificadores.

This 'reclassification of classifiers', beyond the misclassification outlined earlier, represents a politically salient attempt to appropriate a term – proposed by the Uruguayan waste-picker movement and its allies as a way of respectfully referring to all practitioners of the trade (Clara 2012) – to grant recognition and legitimacy to a small subsection working within the state. By stripping other waste-pickers of their legitimacy, these can more easily be dispossessed of their materials and livelihood. Unlike Samson's South African example, where all waste-pickers were denigrated as scavengers in order to justify their neoliberal dispossession and exploitation by private capital, in Uruguay a certain section of clasificadores are selected to be socially included and slotted into a socio-religious narrative of increasing rights, recognition, and dignity (IM 2015). Yet as a result, a large majority found themselves in more precarious circumstances.

Conclusion

In this chapter, I have drawn out the dynamics of some clasificadores' transition from the informal to the formal economy through a focus on care, (mis)classification, and enclosure. I have suggested that, partly due to the legacy of Catholic social work, clasificadores were identified by the state as vulnerable workers whose jobs in Montevideo's recycling plants were 'protected' to a much greater degree than in other formal sector workplaces. For *acompañantes* like María del Carmen, formalisation meant the provision of long overdue access to the benefits of work-based citizenship, while for NGO coordinator Richard, social inclusion 'settled a debt between clasificadores and the Uruguayan society that had marginalised them'. Yet socio-psychological care and educational opportunity could only be provided by the paternalistic state to those workers who embraced certain subjective frames. Sitting in a classroom, accepting socio-psychological support, and embracing the 'vulnerabilisation' of their bodies proved particularly challenging for many clasificador men, whose identities were based around resilience, autonomy, and strength. The codification and formalisation of the criteria of vulnerability needed to access waste meant that certain affordances of the commons were lost as waste's status as a municipal good was reinforced through its enclosure in 'anti-vandal' containers, diversion from the landfill, and delivery only to those who embraced wage labour and new forms of environmental, labour-based citizenship.

I want to end this chapter with a story that brings together its central themes. As I have suggested, the reconfigured gender dynamics of the plant meant that women gained prominence as representatives, while the reconfigured political dynamics entailed a move away from a confrontational relationship with police at the landfill to a more nuanced relationship with the state, where hostility to certain forms of authority fused with elements of a new paternalism. Unlike the waste commons, which clasificadores accessed with few conditions, work at the plant involved entanglement in a citizenship project of rights and responsibilities, meaning that excessive masculinity had to be curbed. But the changes also brought about the possibility of making new demands on the state and invoking forms of worker mobilisation of a kind that will be explored in the following chapter. Towards the end of my fieldwork period, Aries plant workers went on strike, demanding a pay rise and the provision of Christmas hampers for workers and their families.

In asking for the hampers, clasificadores sought to continue a gift exchange that took place not only with intermediaries (such as Álvaro's intermediary father, who gave Gorda Bea and other 'clients' wine and pan dulce every Christmas) but also with Catholic *acompañantes*. When the

Intendencia acceded, they provided festive treats like fizzy wine, nougat, and pan dulce in orange bags that were left over from a previous unsuccessful scheme for the household separation of a recyclable fraction. The 'orange bag' scheme had been roundly mocked by the media and observers because of the failings of an infrastructure of enclosure: the lack of provision of suitable street-level containers had meant that neighbours simply put the orange bags into normal bins whose contents were disposed of in Felipe Cardoso. Now, these leftovers from a failed municipal classification scheme were given a second life as a result of the classification of waste-pickers as vulnerable workers, their social inclusion in recycling plants, and the resultant increase in their bargaining power. A material infrastructure designed to contain waste was reprogrammed to contain labour unrest in a manifestation of at least one form of *requeche* consumption that authorities found acceptable.

5

Precarious Labour Organising and 'Urban Alambramiento*'*[1]

Patricia Gutiérrez, known as La Pato, long-time secretary of the Uruguayan waste-pickers' union, is draped in the Uruguayan flag, and chained up outside the offices of the Intendencia. Now in her fifties, La Pato is a respected community and union leader. Together with a group of fellow clasificadores, she is protesting the rolling out of the sealed containers that prevent street waste-pickers from accessing what is inside, the confiscation of waste-pickers' horses, and the pay and conditions for clasificadores employed in newly created recycling plants, in which Pato herself has started working. With dark, indigenous features, an ever-present smile, and a forceful manner, she is a lifelong waste-picker, a profession she has chosen, she insists, rather than being forced to do. She is a classifier, she clarifies, not a 'rummager', an activity that she associates with rats and pigs. She likes her profession, feeling that it suits her predilection for freedom and autonomy. Her cart, and the discarded materials that she has collected in it, have enabled her to raise four biological and six adopted children. Now, she is worried that, with the closure of streets and enclosure of waste, other clasificadores will be denied that chance. 'The struggle is for clasificadores to be able to continue working,' she tells me. 'If we had to invent this job ... who are those from above, those who govern, to tell us that we don't have the right to work?' While citizens across a Europe riven by austerity are urged to 'reinvent themselves as entrepreneurs' (Narotsky 2020: 11), waste-pickers in Uruguay and across the Global South, who to a large degree reinvented recycling in their respective terrains, find their 'conquests' at risk.

In this chapter, I explore the distinguishing features of waste-picker labour organising through a focus on the clasificador trade union, the UCRUS in order to gain a greater understanding of how clasificadores organised against enclosure. For the most part lacking traditional union tactics such as striking, how did the UCRUS attempt to pressure or leverage its antagonists? How did it organise such a dispersed sector, and which of its constituency's issues did it have to mediate? As we shall go on to see in this chapter, a large part of the UCRUS's struggle during 2014 was against and in the context of particular forms of enclosure. First, there was the enclosure signified by the rolling out of waste containers that were designed to prevent the extraction of materials by clasificadores in the street. Second, the union struggled against the prohibition on clasificadores circulating in certain areas of the city, which were effectively closed off to them. Third, a series of laws that prevented businesses from donating their wastes to informal sector actors such as clasificadores represented a form of legal enclosure. Finally, the union negotiated around, though did not necessarily oppose, the forms of enclosure in recycling plants we have already discussed.

Beyond enclosure, this chapter also speaks to wider challenges for unions attempting to organise in the informal sector, such as the lack of an immediate employer, the representation of informal and formal workers in the same trade sector, and demands that may be related to rights and spatial issues rather than pay claims. In an important way, the demands of clasificadores and their union involve fighting for the right to circulate in the city, and their case thus connects with a renewed interest in various 'rights to the city' (Harvey 2008, 2012; Lefebvre 1968). Instead of constituting an example of a fight to *receive* a service (such as electricity, water, etc.), the clasificadores' struggle includes the demand to be able to *provide* one, in the face of state attempts at regulation, formalisation, and dispossession.

The first section of the chapter offers a brief history of the UCRUS, highlighting how the regional circulation of recycling activists shaped its programme and ideas, and how the two most important events in its history involved struggles against enclosure. I then turn to circulation as a demand that is also enacted performatively as a tactic. The subsequent section looks at efforts to close off streets to recycling workers who navigate by horse and cart (*carreros*) in the city by sedentarising them into recycling plants. An exploration of the links that clasificadores make to the historic struggle of the gaucho against the nineteenth-century enclosure of Uruguay's countryside, known as alambramiento, leads to the conceptualisation of processes of contemporary 'urban alambramiento'. Finally, the chapter turns to the demands placed on UCRUS activists to circulate throughout the city, comparing this 'mobile unionism' with the 'sedentary unionism' of the trade union congress Plenario Intersindical de Trabajadores–Convención Nacional de Trabajadores (PIT-CNT) of which the

UCRUS forms a part. The conclusion suggests the relevance of the case of the UCRUS to research on a wider 'unconventional' labour movement.

The UCRUS: a brief history

According to the PIT-CNT delegate who coordinated with clasificadores, the UCRUS was a 'very peculiar' union. It did not deal with a direct employer but rather made its claims to public bodies, the Intendencia in particular. Clasificadores were not municipal employees, however, and thus lacked a range of tactics available to public workers, not least the withdrawal of their labour. Aside from this, the clasificador labour force was precarious, dispersed, and fluctuating, with many clasificadores coming in and out of the activity as they obtained short-term odd jobs. The clasificador was, for the delegate, something of an 'abstract entity', with its shifting population comparable to those of seasonal industries like fishing or farm work. Yet despite the peculiarities of the sector, the UCRUS boasted of being the only recyclers' organisation anywhere in the world to be affiliated to a national trade union federation, as La Pato explained:

> Fifteen years ago, we started to organise ourselves into a union, something that we clasificadores had never dreamed of, that one day we might have our own union ... do you know that we are the only [waste-picker] union in the entire world affiliated to the national trade union federation? We're proud to be affiliated to the PIT-CNT, and to be considered like any other worker.[2]

The history of the collective organising of Uruguayan clasificadores can be divided into two interrelated areas: the development of the trade union and the formation of cooperatives. In the 1970s, clasificadores involved in the waste-paper trade had been subject to severe police repression in the city's Ciudad Vieja. Not only were their carts burnt and horses confiscated, they were also evicted from central urban spaces where they were either living or storing their materials. An indication of some form of organisation among these workers was that they sent a delegation to the press to highlight the importance of their activity to the economy (*El Diario* 1974). In the late 1990s, a clasificador of the older generation who had once been victimised by the dictatorship attempted to organise clasificadores into a union, with limited success (personal communication August 2014). Another key predecessor of the formation of UCRUS was the Organización San Vicente, a territory-based advocacy group linked to the radical priest Padre Cacho that helped make clasificadores visible and still fights for improvements in their treatment, sanitation, and living conditions. As we have seen, other moments of intense repression have been reported in the press, but contrary to this, organised resistance or opposition from clasificadores is not usually documented (although this does not mean that it did not occur).

As for the UCRUS, its story is intertwined with that of El Viejo, its long-time adviser and co-founder. A former fisherman and political militant, he had begun working with a group of *carreros*, promoting their cooperation so that they might achieve better prices selling their materials collectively. They held their foundational assembly in April 2002 and were able to send a delegation to May Day for the first time that year. The union was strengthened when the group was contacted by clasificadores at Felipe Cardoso toward the end of 2002. The landfill had long attracted the largest concentration of clasificadores in the city and so they made up an important constituency for the incipient union. When police attempted to enforce a new fenced enclosure of the landfill in 2002, a group of clasificadores decided to chain themselves across the road from the entrance. El Viejo and others arrived and suggested that they cross to the gate itself to prevent the entry of municipal trucks. This confluence at the dump of *gateadores* like La Gorda Bea with El Viejo and the *carreros* with whom he had been working is the most important foundational myth of the UCRUS. El Viejo described it this way:

> We didn't say, 'Cross [the street],' we told them, 'Let's cross together.' Because it was predictable that the police wouldn't like it. And that we'd have problems. But let's take the risk and cross because we understand that it's the only way. Two hours later, word came from the Palacio [city legislature]: 'We'll negotiate.' So, [it was] a triumph … as you can imagine, an exceptional triumph.

The power that garbage workers' strikes hold over municipal governments has long been recognised: from the Memphis garbage strike of 1968 (Collins 1988) and the refuse strike of the British Winter of Discontent of 1978–9 (Martín López 2014) to the recent strike of Rio de Janeiro garbage workers timed to coincide with the World Cup. In a vulgar application of the literature on the metabolism of the city (Heynen et al. 2006; Wolman 1965), blocking access to the dump effectively meant shutting off the city-subject's only toilet, refusing to let it relieve itself. Negotiations were successful and clasificadores were granted access to an internal road that lay in between two dumps. There, 30 trucks specially selected for the quality of their material would be dumped and shared by around 150 clasificadores. This marked, in the words of El Viejo, the founding (or refounding) of the UCRUS, and it was an event to which he referred back frequently, if not excessively. Uruguayans are said to be particularly nostalgic, and in thinking back constantly to this event, one militant spoke of El Viejo '*maracanando*' – that is, constantly referring back to old triumphs, as in the case of Uruguay's 1950 World Cup victory in Brazil's Maracanã stadium. After a period of several years, workers at the landfill were moved to an improved site. In their first steps toward cooperativisation, they adopted the provisional name Cooperativa El Viejo for what would become COFECA, in honour of the man who helped them win that crucial first fight and consistently accompanied them through later, more complex struggles.

The UCRUS, meanwhile, began to hold regular meetings, first in the Chemical Workers' Union premises and then in a community centre (the Galpón de Corrales) in front of a worker-recovered tyre factory, FUNSA, in the working-class Villa Española neighbourhood. It developed links with a series of institutions, such as the public University of the Republic's large outreach programme, and gained entry to the PIT-CNT as a member 'with a voice but without a vote', because they never had the 300 registered and paid-up members required for this. The union gained legal status and cemented the positions of president, secretary, and treasurer. It also attracted a diverse group of sympathetic *técnicos* (technical advisers) and *asesores* (advisers): from students and professionals to far-left political activists.

The example of the Brazilian *catadores* movement was important for the Uruguayans. In 2003, a delegation of over thirty Uruguayan clasificadores travelled to Caxias do Sul for the founding conference of the Red-LACRE (the Red Latinoamericana de Recicladores, or Latin American Recyclers' Network). It was then, and in the second conference of 2005, that clasificadores like Nico Trastos and the activists who travelled with them were able to observe the impressive infrastructure that the *catador* cooperative movement in Brazil had managed to establish. This consisted of large recycling plants with technologies that could wash, press, and shred materials, as well as the democratic and organisational infrastructure of collective decision making in formal cooperatives. The UCRUS took inspiration from its Brazilian counterpart, which presented itself not as a union but as a movement: the Movimento Nacional dos Catadores de Materiais Recicláveis (MNCR). Both the UCRUS and the *técnicos*, some of whom would later work with clasificadores as part of the MIDES, adopted key aspects of the MNCR's programme. This included pre-classification of recyclables by citizens (known as *separación en orígen*) and the establishment of recycling plants managed by *catadores*/clasificadores in a cooperative fashion. Such activity represents the crucial circulation of ideas and experiences throughout the sector at a regional level.

To synthesise its history during the following years, the UCRUS attempted to represent *carreros*, as well as a growing number who started to form cooperatives, following a model promoted by MIDES and inspired by neighbouring countries. However, a peculiar dynamic arose whereby a majority of clasificadores continued working individually or in family units with horses and carts, while those who opted for cooperatives were joined by political militants, principally from anarchist and Trotskyite traditions, some of whom were former industrial workers with experience in trade union organising. Similar dynamics have been observed in other contexts, such as that of former Bolivian mine workers in the Cocalero movement (Grisaffi 2019) and in neighbourhood organisations in cities such as El Alto (Lazar 2008). In Uruguay, anarchists, many of them Italian immigrant tradesmen, played an important role in the development of trade unionism; the first Uruguayan trade union federation, the Federación Obrera Regional Uruguaya (FORU, 1905–23), was explicitly anarchist in orientation (Errandonea and Costabile 1968; González Sierra 1989). Anarchist activists within the UCRUS continued a national tradition of labour organising that favoured direct action over negotiation, with these political militants-turned-clasificadores becoming some of the union's most active members.

While prospective cooperativists were courted by the MIDES and other *técnicos*, and given training sessions and limited infrastructural support, *carreros* experienced what union officials described as an alternation between repression and tolerance from the municipal authorities and the police. As we have seen, the activity of clasificadores collecting recyclable public waste in the streets had been decriminalised after Tabaré Vázquez was first elected as mayor of Montevideo in 1990. From the beginning of the 1990s, the Intendencia carried out surveys of clasificadores and began to regulate their activity, distributing cards and registration plates that allowed them to circulate in the city, while simultaneously prohibiting the activity of those without them. Certain arteries of the city were closed to all clasificadores, such as the principal city centre thoroughfare Avenida 18 de Julio and the coastal road known as La Rambla. Being banned from these streets was not hugely disadvantageous for clasificadores, and it was generally accepted that it was necessary for traffic purposes (the Rambla in particular is high-speed and often congested). The denial of access to entire neighbourhoods, however – so-called exclusion zones – was different. As a pilot scheme, the municipal government set up a system of domestic waste classification with corresponding bins in these areas; the recyclables were picked up by the municipal government and taken to clasificador cooperatives, while *carreros* were dispossessed, anticipating the later model of the Ley de Envases plants.

Alongside restrictions on circulation, it was often the municipal confiscation of horses and carts that most challenged the clasificadores' livelihoods and brought them onto the streets. The confiscation was justified with reference to supposed traffic infringements or mistreatment of horses, but clasificadores claimed that the process was unjust and arbitrary. It was a period of severe repression in 2008, during Ricardo Ehrlich's term as mayor, that led to the march that became the other UCRUS event El Viejo constantly celebrated:

> The famous 13th of February, 2008, arrived. Faced with the indiscriminate confiscation [of horses and carts], a demonstration took place which was so important that ... the repressive impetus was extinguished and [clasificadores] could continue working in the streets, within certain limits. The march of February 13th was a very significant march since it drove home the rights of the clasificadores and [the idea] that they had a legitimate function.

The two key moments in the history of the UCRUS thus responded to clear forms of enclosure: the first a renewed attempt to enclose the landfill, the second the enclosure of parts of the city to clasificadores and the confiscation of their horses when they continued to circulate. This is one of the characteristics of union struggle that carried over into my research period, as I describe in the following account of another 'march of the carts'.

The UCRUS in action: 'The nation was built on horseback'

When I arrived back in Montevideo at the end of 2013, clasificadores were on the march again. Or rather on the trot. I met them in front of Uruguay's legislature, where they had mobilised mostly with horses and carts, travelling from different points throughout the city. The march was a family affair, including young children, adolescents, parents, and grandparents. But the mood was angry. They were protesting a decree proposed by a Frente Amplio councillor in the regional council (Junta Departamental), which would prohibit clasificadores from entering the Ciudad Vieja to collect waste. There were already several central areas of the city in which the clasificadores could not circulate with horse and cart, and a decree penalised businesses that gave their waste to informal sector clasificadores rather than formally registered waste contractors.

The Ciudad Vieja prohibition was justified with reference to the narrow, historic streets and the long-time complaint that clasificadores were an eyesore and left rubbish in the streets after rummaging through containers. The area generated valuable waste – particularly its scores of office blocks that disposed of prized white paper. The proposed regulation did not distinguish between clasificadores who had regular pick-ups from offices and those who might pass through on foot to search bins in the hope of finding food, paper, or something else of value. Thus, many clasificadores found their quite lucrative routes and established relations with clients in jeopardy. Unlike other cases where property speculation and neoliberal governance combine to evict the poor from urban living spaces (Barbero 2015; Hertzfeld 2010), here clasificadores were threatened with eviction from labour circuits and dispossessed of valuable 'surplus material'.

From the legislature, the clasificadores marched to the Junta Departamental. They cut through the main arteries of the city, blocking traffic, until they reached the narrow streets of the Ciudad Vieja, where the council building is situated. The horses and carts bore diverse and colourful political slogans. On the cart of the UCRUS president was scrawled a simple quote from Uruguay's founding father, José Artigas: 'The cause of the people does not admit the slightest delay.' Flags of Uruguay and the tricolor of Artigas, also used by the Frente Amplio, abounded. On a bicycle and cart one man had written, 'We're fighting for our jobs.' Another sign read, 'My struggle for bread and work continues.' On one cart, alongside drawings of horses, was written, 'The nation was built on horseback.' Most protesters travelled by horse and cart, while others trotted along. The carts ranged from the simple and workmanlike to the elaborate and ceremonial that bore impressive, brass-decorated reins that clasificadores use for special occasions. The younger men were dressed in a popular urban sports style known as *plancha*; most wore Nike baseball caps. Some combined this with hair worn long, as is popular in the more peri-urban shantytowns. The march represented the periphery come to visit the centre, whose suited businessmen looked on, aghast or bemused. Some speeches were made from horseback. One long-haired and bearded activist shouted that clasificadores and their horses had 'the same right to circulate in the city as anyone else in their cars'.

While some clasificadores stayed to look after the horses, others streamed inside the regional council building. There was little security or police presence, and the young men in particular seemed hyped-up and excited, knocking things over and pushing into offices. Some councillors rushed for cover, and when the Frente Amplio council leader came out to speak to the demonstrators, he was quickly surrounded. Clearly frazzled and shouting to be heard, he told demonstrators that the doors of the council were open but only for those who came to 'dialogue with order and respect'. La Pato and another leader from the UCRUS put forward their demands, but other clasificadores complained that they didn't want 'the same representatives as always'. Another denounced them, saying that the UCRUS 'wanted to send them to recycling plants'. One young anarchist activist was extremely agitated, encouraging others to 'smash it all up!', while another angry young man outside shouted insistently, 'I want to work [*Quiero laburar*]!'

The UCRUS leaders decided that they should march to the Intendencia, the imposing building that houses the municipal council, situated on central Montevideo's main thoroughfare, Avenida 18 de Julio. One older protester, Juan, invited me onto a cart with his son. As we trotted down the street, he told me about his years of classifying and how he always tried to attend demonstrations by the UCRUS. The cavalcade came to a stop outside of the Intendencia, where leaders demanded a meeting with Montevideo's mayor. In the end it was decided that María Sara Ribeiro, head of social policy and right-hand woman of the mayor, would receive a delegation of female clasificadoras. The group started to disperse, but not before something of a dirty protest occurred: the horses, stopped for a long time outside the Intendencia, inevitably began to relieve themselves under the gaze of municipal authorities.

The (en)closure of urban areas and the threat/opportunity of formalisation have been an important part of the struggles of other informal sector workers, street vendors in particular (Hansen et al. 2013; Lazar 2008). Indeed, clasificadores were not the only group to have their circulation throughout the city impeded by the municipal authorities in Montevideo during 2014. Those who clean car windows at traffic lights, another group whose status as workers was in doubt, also faced a new threat to their circulation in the city. In a move that surprised some representatives of the Frente Amplio, the Minister of the Interior and the Chief of Police had moved jointly to prohibit the activity, citing the obscure Article 543 of the Municipal Code which establishes when a pedestrian can circulate on the road. Traffic-light jugglers were also included in this, with the justification that many used such activity as a cover for robbery. During my fieldwork period, the UCRUS tried to make common cause with these and other informal sector workers (street vendors in particular) who were under pressure to sedentarise and formalise their activity, launching an ultimately unsuccessful attempt to establish an Informal Workers' Federation.

Yet if the question of circulation was key to the struggles of other informal sector workers, it was primarily clasificadores who enacted circulation as a *method* of protest. Blocking the entry of waste to the municipal landfill had been a successful tactic in 2002, but it had rarely been used by the union or *gateadores* since then – although it was often threatened. After the successful march of 2008, however, horse-and-cart cavalcades became the tactic most associated with the UCRUS, which organised them several times a year. They were moments when territorially divided clasificadores who worked alone or in small family groups could come together in defence of their common interests. On an ordinary working day, the clasificador on horse and cart would be alone, awkwardly making their way through a sea of cars that would speed past, often with drivers complaining loudly. Members of the public might harass the clasificador about the condition of their horse or the fact that the animal was in the city at all. They would run the risk of an inspector or policeman confiscating the animal and the cart. And were there to be a traffic accident, *carreros* could rest assured that the blame would be attributed to them, whatever the circumstances, as newspaper headlines made clear every time such an accident occurred. Some members of the public were of course friendly to clasificadores, and they had their regular clients who would greet them. But as Pablo, a former *carrero* who now works at the Planta Aries, told me, 'There's always someone who will shit on your day.'

On march days, in contrast, clasificadores had safety in numbers among colleagues, family, and friends. The street belonged to them, and their right to circulate on it was uncontested. The cries of clasificadores as they drove their animals down the normally prohibited Avenida 18 de Julio were joyous. Should any motorist attempt to drive through their march or abuse them, the united clasificadores could easily confront the aggressor, as I witnessed on several occasions. This was a moment where the 'imagined community' (Anderson 1983) of clasificadores could be realised and their numbers complemented by the spectacle of horse-drawn transport, with carts of amazing workmanship. The tactic they used to demand the right to circulate was thus itself a rebellious circulation (Figure 11).

During the course of my 2013–14 fieldwork, the UCRUS decided to complement marches with another form of protest – that of the roadblock – belatedly adopting the method associated with the *piquetero* movement of unemployed workers in post 2001-crisis Argentina (Svampa 2004). If streets were to be closed to clasificadores and their valuable waste enclosed, then they would make sure that they were closed off to the general public too. The UCRUS had attempted this tactic in winter 2013 and had then decided to carry out three roadblocks on different days in different parts of the city, to build up momentum for their *marcha de carros* of September 2014. The success of this tactic was variable in 2014, with one roadblock attended by few clasificadores and much media; another by many clasificadores but no media; and a third only by myself and another research student! Regardless of attendance, however, the roadblocks and marches had the effect of imposing upon the public forms of closure and enclosure that *carreros* experienced on a daily basis.

Figure 11 *Carreros* on a 2013 UCRUS march protesting at the Junta Departamental
Source: Author photo 9 December 2013.

Sedentarisation and alambramiento

The clasificador with horse and cart travelling through the city maps neatly on to the Deleuzian idea of an assemblage. The philosopher used a similar example, that of the knight, to illustrate the concept, wherein the stirrup binds together a 'man–animal symbiosis, a new assemblage of war' (Deleuze and Parnet 1987). In the Uruguayan case, the components of the assemblage were bound together in the word *carrito* (little cart), commonly used to refer both to the clasificadores *and* their mode of transport (as in 'the *carrito* problem'). A common complaint from the public and institutions focused on the circulation of the horse and cart in the city, either because they disrupted transport or because the city was supposedly not the natural environment of the horse. Uruguay's vocal humane society, the Protectora de Animales, was a principal adversary of the UCRUS, and the union accused the organisation of stealing clasificadores' horses, which were then difficult to trace.[3] Yet the assemblage nature of the *carritos* meant that there was always a slight ambiguity over whether the objection was against the circulation of animals in the city or of the poor. There is no doubt that some saw the circulation of poor people in dirty clothes, riding horses and rummaging in bins, as an eyesore for tourists and an affront to Montevideo's infrastructural modernity. A plan to convert some *carreros* into horse-and-cart tourist guides for the Ciudad Vieja had not been well received by the UCRUS. Another scheme to provide motorised tricycles (*moto-carros*) received much press coverage but was initially limited in scale. A more ambitious plan, however, was the relocation of *carreros* to the Ley de Envases recycling plants.

The relocation of *carreros* to plants, whether cooperative or municipal, is as much one of sedentarisation as it is of collectivisation. Mobility and autonomy are important values and practices for *carreros*, as are not having a boss and setting one's own hours. As La Pato explained after she had been recruited to a recycling plant: 'I don't like being bossed around. I don't like having fixed working hours. I don't know how many days I will last in a plant. I am independent, free, as free as a bird.' La Pato also told me that she had always been a 'clasificador by choice, not by necessity'. She explained that when starting out as a teenager, the other options available to her were either low-paid work as a domestic servant, prostitution, or theft. Many of the *carreros* I interviewed preferred the independence of classifying to either a life of crime or the kind of low-paid security work available to the men, and the cleaning or domestic work available to the women, where they would work long hours for low pay. At least on the horse and cart they were their own masters, had control over their working day, and were sure to bring home not only money, but *requeche* as well.

Some UCRUS militants described the *carrero* as a modern descendant of the gaucho, the somewhat mythical nomadic horseman who roamed Argentina and Uruguay until the land was fenced off and enclosed into private ranches in the late nineteenth century (Ras 1996). This process was tantamount to Uruguay's 'enclosure', known as alambramiento, or 'wire-fencing', after the wire that was imported duty-free in order to accomplish the task. Between 1872 and 1882, 64 per cent of the surface of the country was fenced in, with important social and economic consequences (Millot and Bertino 1996: 61). Given the space dedicated to the English enclosures from which this book derives its conceptual understanding of commons and enclosure, it is worth focusing briefly on the Uruguayan process here, which had different characteristics but some of the same effects, such as impoverishment, increased inequality, the strengthening of private property, and improved rural productivity.

One crucial difference between the land enclosed in Uruguay and much of that enclosed in England was that the former was not generally thought of nor is regarded by historians as commons, even if the ambiguous status of land prior to alambramiento means that it may have functioned as such in certain places. As we have noted, cattle have long been central to Uruguay's economy. Prior to independence, Uruguay was briefly joined with Argentina in the United Provinces of the Río de la Plata and the sixteenth- and early seventeenth-century governor of the Río de la Plata, known as Hernandarias, is credited with introducing cattle to Uruguay in the early seventeenth century, the original herd giving rise to a vast, wild colony. This was treated as a limitless resource by raiding parties from Buenos Aires who, primarily interested in leather, often left the rest of the animal to rot (Solari 1953: 315). In such circumstances, the cattle themselves, rather than the land, seemed to effectively function as a commons, and indeed a tragedy of the commons might well have occurred had not restrictions on the slaughter of cows been introduced in a bid to halt a dramatic decrease in the size of the stock (Solari 1953: 316).

Those who raided the 'cattle commons' for leather in the seventeenth and eighteenth centuries were the original gauchos: they traded leather, enjoyed a largely beef-based diet, lived nomadically and autonomously, and gave rise to many of the character traits long associated with the figure (Solari 1953). These include an appreciation of freedom, a disdain for authority, and an egalitarian ethos (as well as a love for certain 'vices', such as gambling, womanising, and violence) (Solari 1953: 177). By the nineteenth century, many of the trends in the transformation of rural work, economy, and land use that would be accentuated by alambramiento were already under way. Recent research (Moraes 2008) has suggested that those disenfranchised by the Uruguayan enclosures were largely *agregados*: rural dwellers living on land that belonged to someone else. *Agregados* might have cultivated a minimal family plot on such land but, unlike evicted crofters in the Scottish Highlands, for example, often dedicated a significant amount of time to the rodeo: the rounding up of free-roaming stock for cattle and land owners.

With enclosure, the need for such labour was greatly diminished and many *agregados*, as well as small landowners, unable to bear the costs of obligatory alambramiento, were evicted from their lands. As Moraes (2008: 92) writes, 'in the short term, this led to the forming of a miserable mass, the founding of rural "rat-towns" [shantytowns]'. Millot and Bertino (1996: 64) give more detail on the treatment of what they call *el gauchaje*, noting that:

> enclosure caused significant unemployment in the countryside.... To leaseholders, sharecroppers, and ruined or expelled small producers (made to leave as much by force as by bankruptcy) were added those from inside the ranch (*estancia*).

Contemporary estimates placed rural unemployment at 10 per cent, or 8,000 families. Cattle rustling increased and, following lobbying from large landowners, the dictatorship of Lorenzo Latorre (1876–80) introduced a Rural Code and public order acts targeting vagrancy. As Millot and Bertino point out:

> the establishment of order in the countryside coincided with the application of the Rural Code and alambramiento: thus a 'vagrant' [*maleante*] was often someone dispossessed of their small parcel of land because of the 'rounding off' of the estates or a farmhand expelled from the ranches and unable to adapt to the new discipline [of waged labour]. (1996: 65)

Millot and Bertino conclude that 'the dispossession of the means of production [cattle] from a part of the population constituted a phenomenon of primitive accumulation that had been gathering steam since the end of the 18th century' and included 'modifications in the rural mode of reproduction and its forms of social consciousness, modelling a [new] man who had to sell his labour force in order to subsist' (1996: 65). Many dispossessed rural dwellers who were not jailed or forced to join the army, police, or paramilitary landowner units, eventually made their way to Montevideo, where some found work in the nascent industrial sector. The parallels with the enclosure of the English commons and the provision of a labour force for early capitalism are thus evident. Yet the parallels with the modern clasificadores should also be clear, where the various forms of waste enclosure, from physical wire-fencing around the landfill, to prohibitions on entering neighbourhoods and the installation of 'anti-poor' containers, can be considered forms of 'urban alambramiento'.

Not all rural workers moved to the city of course, and the gaucho found a modern descendant in the countryside as well, complete with traditional attire and a stronger heritage of horsemanship. During one march, the UCRUS was lent the support of a few horsemen by a rural organisation, and they became celebrities of the march. Rural horsemen and urban clasificadores came together every year at *El Rusbel*, the annual rodeo and horse show that took place at Parque Roosevelt, on the outskirts of Montevideo. In many ways clasificadores could trace just as great a cultural and political heritage to the oppressed and marginalised gaucho, who was often treated as a criminal. Sometimes, they could even claim direct genealogical connections, since many clasificadores were, or were descended from, rural migrants who had escaped poverty and rural 'rat-towns' only to find themselves living in urban *asentamientos* and making a living from waste. As I have argued, many of those affected by the enclosure of waste have already been victims of prior rounds of enclosure of land in the countryside.

Like the gaucho, many *carreros* were also faced with the choice of sedentarisation or elimination. A senior figure at the Intendencia, who was responsible for the municipal management of the recycling plants, argued confidently:

> The clasificadores that are in the street today ... will they be able to continue classifying materials? No, not really. Those who sign up for recycling plants will, but those who don't, will not. Alongside the intermediaries, the profession of informal classification will disappear.

Yet the presence of horses and carts on the streets of Montevideo is surprisingly persistent. Partly, this is due to the geographic characteristics of Montevideo, where the distances between centre and periphery are not so large as to make them unfeasible for horse and cart. Some clasificadores thought that the long-held municipal aim of eradicating the carts was a pipe dream and spoke about how, during the dictatorship, police had burnt their carts in the streets but still had not defeated them. Others were more worried, however, and suggested that while the military had tried a blanket approach, Mayor Ana Olivera was being more 'sneaky' by restricting circulation in different areas progressively, rechannelling the circulation of waste so that *carreros* could not access it and, as I have argued, instituting a form of 'dispossession-by-differentiation'. I now turn to the effectiveness of the union in challenging this differentiation, a form of work I characterise as 'mobile unionism'.

Mobile unionism

During the course of 2014, the UCRUS's weekly meetings in the working-class district of Villa Española that I sat in on were particularly poorly attended, and union officials spoke about holding assemblies in different neighbourhoods in order to attract more people. The city's informal settlements, where most clasificadores lived, were spread out across the city. The organisation of clasificadores by canton was a long-held aim of the union (Fernández 2007: 90), and in the past, neighbourhood meetings had been a common practice, as a disaffected UCRUS founder and former president complained to me:

> When I was in the UCRUS, we didn't have meetings there [in the Galpón]. We went around the neighbourhoods. Every two weeks we went to a different neighbourhood. The place changed. Because … the clasificador can't go so far. If you go to the neighbourhood [*barrio*] … you're in their territory.… They need to come to the neighbourhoods because the majority of the *carreros* are in our neighbourhoods. And they don't come!

In 2014, perhaps twenty or more clasificadores attended meetings at the Galpón, but this was almost always a single visit. They failed to return, either discouraged by the meetings or because they were unable to attend regularly. I was also present on a few occasions when the UCRUS attempted, with mixed results, to circulate in the neighbourhoods. They held a series of neighbourhood assemblies, starting in Marconi. The nature of this first assembly was indicative of the crisis into which the UCRUS had slumped, since this should have been a good place to meet. It is an area with a large number of clasificadores, where Padre Cacho had worked, where the UCRUS had held its foundational meeting, and where the then UCRUS president Cayo lived. Yet aside from a couple of 'characters' who seemed to spend most evenings in the plaza anyway, only a few clasificadores appeared, and union officials were even heckled. The poor quality of the sound equipment meant that it was difficult to make any sense of the president's rambling speech. As if to highlight the lack of interest from *gateadores* in the event, my neighbour Juan, who had previously travelled to Argentina with the union back in 2003, gave me a lift on his scooter, engaged in a frosty exchange with El Viejo, then left me to it.

On another occasion, we headed out to distribute leaflets for a march in some informal settlements to the east of Montevideo. Clasificadores' homes were usually fairly obvious, as materials were stacked up in their yards and they had horses in ramshackle stables. We also asked around the neighbourhood for people who classified, whereupon we would usually be pointed in the direction of someone or other. This got a better response than the ill-fated Marconi assembly. It was an occasion when the union officials were able to listen to their members. Had they been affected by exclusion from parts of the city? What did they think of the new recycling plants? How was the market in materials looking? It was also a chance for the union to report back on its activity, encourage clasificadores to attend meetings, and gauge their opinion on possible actions. Clasificadores were generally friendly, committing themselves to organising their *barrios* and attending the next march. Such encounters demonstrated the potential of the union to reach out to the neighbourhoods and circulate in them, but they were the exception rather than the rule.

When I worked at COFECA and they were coordinating with the UCRUS on the eve of their move to the Aries recycling plant, the morning appearances of El Viejo were a sign of both his own vitality and the frailty of the union. The 85-year-old would take a bus or two and then walk from Camino Carrasco the length of Felipe Cardoso to arrive at COFECA. Several times, he appeared alone, and the others complained that it was always El Viejo who came by himself: where was the rest of the union? At another meeting, a clasificador ally turned up with another of his neighbours. He was meant to be meeting the secretariat to talk about organising, but only El Viejo and I were present. This long-time supporter challenged the former as to why there were not more people present after twelve years of the union's existence, and the union stalwart struggled to answer the question.

Of course, it is not fair to accuse the other union leaders of being immobile, and the extremely leaky and noisy but colourful shell of the Galpón de Corrales hardly represented the pinnacle of union opulence or aristocracy. Instead, the union during 2014 was made up of threadbare militants who had to spread themselves thin. No member of the secretariat possessed a car, so most made their way around on foot, by bicycle, on public transport, or in the vehicle of the PIT-CNT representative. They were rarely paid for union work, which they had to combine with a range of other activities (unlike El Abuelo, who was retired). The interim president's time became even more constrained when he accepted a relatively well-paid temporary job as a security guard on a building site. La Pato attempted to combine her union work with a job as a carer and ended up having to resign from the latter because of her commitment to union activity. She suffered from back problems, largely due to a life of hard work classifying on the streets. The circulation of union officials, regarded as key for successfully mobilising their sector, was thus limited by their own precarious economic situations and ill health.

Sedentary unionism

It is perhaps useful at this point to reflect on how different this situation is from that of a traditional, workplace trade union. Militancy and trade union involvement of course depends hugely on the sector and epoch (Gall 2002). Trade unions are in most cases forced to engage in recruitment (Kelly and Heery 1989; Mason and Bain 1991;), but in a regular scenario, a trade union member would approach the union with a complaint or issue. The key demand of the UCRUS, that of the right to circulate freely throughout the city collecting materials and opposing the enclosure of waste, is also clearly different from the world of pay claims, disputes, strikes, and tribunals of the formal trade union. Not only does the UCRUS not have a fixed list of members who pay a subscription – the union is in principle tasked with representing *every* clasificador – it must also go to the clasificador in their neighbourhood. It is resented if it does not go – and even sometimes when it does. While other trade unions have offices and headquarters, UCRUS unionists are expected to travel around the city, conducting a mobile unionism.

The mobile unionism of the UCRUS can be compared with the sedentary unionism of the trade union federation of which they form a part, the PIT-CNT. In referring to the PIT-CNT unionists as sedentary, I do not wish to denigrate them but merely to pick up on one of the key definitions of sedentary: 'tending to spend much time seated' (*OED* 2020b). This characteristic is defining not only of some trade union officials, who inevitably spend much time in meetings, but of millions of other jobs as well – including that of the author! – and should not carry any connotation of inaction. Rather, I wish to contrast the expectations placed on UCRUS militants with those from the formal trade union sector, in which the UCRUS operates rather uncomfortably.

The year 2014 was particularly sedentary for the PIT-CNT. It was an election year, and the links between the union federation and the governing Frente Amplio were becoming increasingly corporatist (Silverman 2011). While UCRUS activists spoke of how they had to put pressure on politicians to make commitments during this period, sources within the PIT-CNT told them that they wouldn't be organising any demonstrations or marches that might challenge the government or upset the chances of a Frente Amplio victory, as a lacklustre presidential campaign from former president Tabaré Vázquez faced a challenge from Luis Alberto Lacalle Pou, the youthful if vacuous son of former neoliberal president Luis Alberto Lacalle.[4] Successive Frente Amplio governments had brought about improvements in workers' rights and conditions, with particular victories being the institution of the eight-hour rural working day and the introduction of legislation that enabled neglectful businesses to be criminally charged for the death or injury of a worker. In addition, nine members of the secretariat of the PIT-CNT featured as candidates for the Frente Amplio on their national ballots in 2014. The only action taken by the union during the election period was an electoral stoppage to declare support for a Frente Amplio government and opposition to a government of the two centre-right parties, all the while declaring its 'class independence'.

When the UCRUS did go to the PIT-CNT's headquarters, as they did for a period during my fieldwork, it was either to attend meetings of the Health, Safety, and Environment committee (under whose banner they fell) or to be heard at the federation's executive committee. The attempts to attend the executive proved rather confusing and exacerbated the insecurities of the UCRUS with respect to its relation to the federation. The UCRUS asked the committee to put out a statement supporting them on three key points. The first was that it publicly oppose plans for the incineration of waste; the second that it oppose the decree that barred *carreros* from collecting commercial waste; and the third that it support a plan for a levy on plastic bags, which would pay dignified salaries to clasificadores at a proposed larger recycling plant. After much confusion and waiting around, the PIT-CNT's eventual response was to arrange meetings with both the presidency and the Intendencia to discuss these issues, while quietly voicing support.

The way in which the UCRUS perceived their reception by the PIT-CNT signified not only the precariousness of the UCRUS leadership but also that of their link with the PIT-CNT, despite the pride they took in it. During the course of my fieldwork, UCRUS militants were uncomfortable with sedentary union practices, such as negotiations with the Intendencia, that were secured after the September march. During four sets of tripartite meetings between the Intendencia, UCRUS, and PIT-CNT, the union failed to make any headway on the seven-point programme they took into the meetings, and there was soon talk of more demonstrations, marches, and hunger strikes.

The comparison between the UCRUS and the wider PIT-CNT can be set in the historical context of Uruguayan trade unionism. In the early twentieth century, militant anarchist trade unions predominated (most within the FORU), in what Errandonea and Costabile (1968) classify as an 'oppositional model' of trade unionism, where militant strike action was often the first option. In contrast, these authors chart the slow emergence of 'dualist' trade unions, which might have a revolutionary or radical leadership but engage with more immediate concerns of the workers. For these, strike action was a last resort (see the dual role of unions in Darlington 2013: 113). Although there are clear differences in the sector represented (the FORU mostly represented skilled artisans with significant leverage), there are also clear continuities between the FORU and the UCRUS's preference for direct action over negotiation. Unions affiliated to the FORU went somewhat further than the UCRUS in the militancy of their actions, nonetheless: in the 1922 municipal waste-workers' strike, for example, strikers put bombs in bins to discourage strike-breakers from collecting them, leaving two dead as a result (Errandonea and Costabile 1968: 115).

'We do recycle, the containers do not!' – international parallels

To what extent are the kinds of struggles that the UCRUS have been involved in – against the enclosure of containers and the closure of streets – and the methods used to achieve their aims – mobile unionism and circulation – replicated in the struggles of waste-pickers elsewhere? As I have suggested, the circulation of ideas between Uruguayan, Brazilian, and Argentine waste-picker groups was common, as were exchanges with the other waste-pickers represented in the Red-LACRE network. While this did not mean that their issues and strategies were the same, there were very often overlaps between them and in this section, I will gesture towards the broader relevance of my focus on enclosure to waste-picker organising.

The two key moments in the history of the UCRUS were protests organised against two forms of enclosure: landfill and street. The modernising and fencing in of sanitary landfills – a key component of what I call hygienic enclosure – is something that waste-pickers regularly have to organise against. Such struggles may even, as in the case of the UCRUS, give birth to organised waste-picker labour movements. I have argued that waste-picking is not necessarily an individual labour – it is very often carried out by families or partners – but it is certainly true that more organisation and unity is needed in the face of threats to a common livelihood than in the exercise of that livelihood on a daily basis. As we noted in the introduction, waste-picker struggles against the enclosure or closure of landfills have been documented across the world, from India to Iran, Egypt to Ethiopia.

Containerisation is a further technology of enclosure. In September 2012, I attended a demonstration organised in Buenos Aires by the Movement of Excluded Workers (MTE), which includes waste-pickers. Unlike the UCRUS, the MTE successfully negotiated with the local government in Buenos Aires to regularise the work of street waste-pickers: in addition to the sale of materials, thousands of *cartoneros* were remunerated with a monthly stipend provided by the municipality. Yet, like their Uruguayan counterparts, they were also angry about the rolling out of new containers by the municipality, in this case bins that could only be accessed by keys distributed to the concierges of city centre apartment blocks. One of the differences with the Uruguayan case was that the law in Buenos Aires was on the side of the *cartoneros*: consecutive decrees passed in the 2000s (Law 992/2002 and Law 1854 or the Zero Waste Law) stipulated not only that the local government needed to recycle waste but also that it should give priority to waste-picker cooperatives in the delivery of this service.

At the rally, amidst pouring rain, *cartonera* leaders poured out angry invective against containers, toppled them over, set fire to them, and emblazoned them with the words 'We *do* recycle, the containers *do not*' in graffiti stencils. A *cartonera* drum band created a vibrant, carnivalesque atmosphere. 'We're here because of the containers', one female waste-picker told a film crew documenting the event, 'it's because of them we are losing our raw materials and are not able to recycle anything: all we want is to work in a dignified way and earn our daily bread' (MTE Argentina 2013). 'We're asking them to remove the bins so that we can work', added another. MTE leader and lawyer Juan Grabois accused the city government of wanting to divert the recyclables to business cronies running new recycling plants. The MTE leader Sergio Sánchez, nicknamed 'the Pope's *cartonero*' because of his close relationship with Pope Francis, former archbishop of Buenos Aires, addressed the crowd, suggesting that the containers worked against social inclusion and all that they had achieved.

In Brazil, I was also invited to travel around with leaders of the MNCR in the Brazilian State of Rio Grande do Sul. As well as campaigning against the closure of a landfill in the border city of Uruguaiana, the more general campaign and concern surrounded the potentially massive threat posed by the installation of incinerators/ waste-to-energy plants. Together, we toured regional consultation meetings about incineration, where the MNCR leaders and I faced off against engineers representing multinational companies. Also with us was Brazilian anthropologist Lolí Wirth, whose work offers a broader perspective on the campaigning priorities of Brazilian waste-pickers. Alex, the MNCR leader who invited me to campaign with his group after we met in Uruguay, and who has recently authored his own book about his experiences, told Wirth that the first mobilisation of the MNCR had ended with waste-pickers collecting and then separating waste in the centre of Porto Alegre, creating the spectacle of a 'mini-recycling plant' (Wirth 2016: 142). The visibility and spectacle of this protest has clear parallels with UCRUS demonstrations and, as Wirth argues, was important in a context where the work of Porto Alegre *catadores* was relatively invisible, hidden away in peripheral recycling plants and lacking the street presence so evident in Montevideo and Buenos Aires. 'Working in the street, in city centres, calling the media and dialoguing with the public would', Wirth explains, 'become a form of protest adopted [by the MNCR] in various cities' (2016: 143).

As in the case of the clasificadores, the demand is the same as the tactic but in this case, these are more explicitly focused on classification than circulation. On another occasion, the Porto Alegre pickers also organised a 'march of the carts', with marches, occupations, and public displays of waste-work forming the three methods of the early phase of *catador* mobilisation (Wirth 2016: 143). Just as La Pato from the UCRUS told me that clasificadores had initially been predisposed to more violent methods of protest before El Viejo advised dialogue, an MNCR leader also told Wirth that initially they had been all about 'adding fuel to the fire' and 'causing trouble', but had then learned to dialogue as equals with those from local government (Wirth 2016: 146). According to Wirth, this marked the start of a dual track phase in which the movement combined 'institutional dialogue' with 'struggle in the streets'. Yet both on the streets and in dialogue with authorities, the MNCR raised the same series of demands and objections: against private sector proposals for 'incineration, containerisation, and mechanical classification' and for 'popular recycling' (*reciclagem popular*) (Wirth 2016: 148).[5]

In the Argentine, Brazilian, and Uruguayan cases of waste-picker collective organising therefore, we find similar campaigns and methods of protest where demands and tactics coalesce, whether these are to do with circulation or classification. Overlapping concerns were due both to common issues faced on the ground, and attempts to learn from each other's struggles through physical face-to-face meetings, as in the visits of MTE leader Sergio Sánchez and MNCR leader Alex Cardoso to Montevideo. Yet one could not help notice that the organisation and influence of the MNCR and the MTE was greater than that of the UCRUS. This cannot be explained by the political orientation of the government interlocutors in each country, since while the national governments of Brazil, Argentina, and Uruguay were, at the time, centre-left progressive, that of Buenos Aires was led by centre-right Mauricio Macri, who would later become President of Argentina. More relevant was the inability of the UCRUS to successfully establish a 'dual track' of organising that combined institutional dialogue and street mobilisation. Yet perhaps it was the UCRUS's predilection for street action in defence of the autonomous *carreros*, those descendants of the *gaucho* struggle against alambramiento, which partly explained why horse-and-cart waste-pickers could still be seen in the streets of Montevideo but no longer in Buenos Aires or Porto Alegre.

Conclusion

The focus on circulation and enclosure in this chapter emerged from the simple demand of *carreros* to be able to circulate in the city, including in the most affluent and waste-rich areas, without harassment. However, it soon became clear that circulation was important for the UCRUS in many other ways as well, from the regional circulation of recycler activists that inspired its programme, to the tactic of blockage and circulation through which it acted out its demands or otherwise pressured the municipal government. The UCRUS had to deal with the municipal rechannelling of waste from poor workers in the informal sector to those recently incorporated into the formal sector while simultaneously attempting to represent all clasificadores, from plant workers and cooperativists to those working at the landfill or with a horse and cart, on a bicycle, and on foot.

It might appear that negotiating the circulation of surplus material, instead of disputing the allocation of surplus value, is peculiar to the struggle of waste-pickers, in Uruguay and beyond. Yet contemporary social movements have also moved beyond struggles over returns to labour and capital and the UCRUS shares with them an attempt to preserve the livelihoods and social reproduction of clasificadores (Narotsky 2020: 13). In struggling against forms of urban enclosure – and to work as well as live in areas like Montevideo's Ciudad Vieja that are targeted for speculation, development, tourism, or gentrification – clasificadores also make common cause with the urban poor worldwide. In place of the negotiation tactics of sedentary unionism, the UCRUS has drawn on the Uruguayan anarchist tradition to advocate direct action, and on national history and myth to create a comparison between their struggle and that of the dispossessed but autonomous gauchos of yore.

Like other post-dictatorship Latin American countries under centre-left governments, the UCRUS had to tackle forms of repression that were insidious but not flagrantly brutal. Clasificador carts were no longer burnt in the street (although they might still be confiscated), but businesses were now penalised for giving their waste to informal sector waste-workers. Rather than having to defend their membership against wholesale criminalisation, the UCRUS had to navigate a complex, evolving situation of dispossession-by-differentiation, where some clasificadores were given 'dignified', if low-paid, jobs in recycling plants, while the unlucky ones were simply dispossessed of their livelihoods.

During the course of the year, it became clear that the union was in difficulty, despite the best efforts of the small team of activists who sustained it and tried to organise the sector. While clasificadores and the UCRUS performed their right to circulate and break out of their enclosure in periodic marches, and there were no reports of *carreros* surrendering their horses on going to work in plants, union informants felt that long-standing efforts to sedentarise the nomadic *carreros* were finally making some headway. The presence of the poor circulating in the centre of the city was being diminished, as has been the case in other Latin American capitals (Medina 2000: 52). As the UCRUS secretary told me, if she wasn't able to access valuable waste in affluent Pocitos, what would she go there for? *To go for a walk?*

Conclusion: Circular Economies, New Enclosures, and Commons Sense

As I have set out in this book, Montevideo's waste recovery infrastructure has until recently been almost completely 'informal' and maintained by clasificador kinship, which at least in my field site was centred around adult clusters rather than exploitative nuclear families. Such kinship structures do not imply the cast-iron loyalties and responsibilities that anthropologists used to ascribe to kinship models like the clan or tribe. Instead, brought up in precarious circumstances with weak or transient father-figures, clasificadores often relied on the care and security embodied in the sibling bond. One way of caring for family members as they moved into adulthood is through the provision of waste and waste-labour. Positioned outside of wage labour, clasificadores are connected to wider society and international markets through a waste trade that they pass on to their brothers, sisters, children, and wider kin networks. Forms of popular knowledge, access to waste spaces, and waste itself thus circulate along familial lines.

The division of the book according to different waste-work locations allows us to take stock of the connectivity of diverse classificatory spaces. The book began with an encounter with clasificadores on horseback who proclaimed that rubbish 'belongs to the poor'. I have used this claim as a foundation for the argument that Montevideo's waste can be considered a commons, taking you on a journey around the Felipe Cardoso landfill and its environs, through classification plants and precarious trade unionism. What links *carreros* with *gateadores* like Juan and Aries workers like Ana Clara? One thing is Uruguayan waste politics, and the Ley de Envases plants designed to negatively impact on street and landfill waste-picking. The provision of employment for a small number of clasificadores has made it easier for the local government to undermine forms of waste-labour that it considers illegitimate, undignified, or irksome. Recyclables deposited by the public in hermetic containers are now redirected to plants, away from kerbside waste-pickers who are also prohibited from circulating in certain areas of the city.

When I returned to my field site in early 2020, the gulf between conditions in the plants and those on the streets had widened. Plant workers had managed to secure a significant increase in their wages, making them more than double what they received during my fieldwork period of 2014. *Carreros*, on the other hand, faced new restrictions on their right to circulate in the city. A new fence installed around the perimeter of the landfill had been dotted with police cabins and *gateadores* received a restricted number of trucks at Usina 5. Some hope could be found, nevertheless, in the central place given to clasificadores by the new Intendente of Montevideo, who in her opening policy speech argued that they should be considered part of the solution rather than the problem. The spaces of the Trastos, meanwhile, were under serious threat, with Nacho and Natalia, as well as La Negra, taken to court by the owner of the land that they had occupied for over thirty years. Customary right and access could still be trumped by property law, and judges within this system had little sympathy for the caring relations that such spaces entailed.

The possibility of claiming the kind of access to things that I have described in this book relies on them having been abandoned and wasted. No matter that such things are then 'unwasted' to become gifts, *requeche*, *material* or commodities, the waste phase is a crucial one in their social life because this is what enables those in a position of vulnerability to gain access without, in many cases, being obliged to give anything back in return. As such, this book puts forth a defence of waste as a category and warns against the dangers of making rubbish disappear through its recategorisation as energy or resource by way of policy initiatives like zero waste or the circular economy. Since research for this book was conducted, the concept of the circular economy in particular has gathered speed, championed by the Ellen MacArthur Foundation (EMF), with its tripartite definition of the circular economy as designing out waste and pollution, keeping products and materials in use, and regenerating natural systems (EMF 2020). Even this short definition, laudable as it sounds, should set alarm bells ringing in readers' ears, given my defence of the waste category as something often integral to the ability of waste-pickers to gain access to materials. Designing out waste, in other words, often involves designing out the waste-picker.

'We're part of the sharing and circular economy!' an enthusiastic and likeable entrepreneur tells me at a London event bringing together researchers, policy makers, and industry figures interested in plastics and the circular economy. His particular scheme involved renting out reusable plastic cups and glasses that would be tracked through an app and could be redeemed for a small cash transfer if returned at a company collection point. The aim – to reduce the use of disposable or single-use cups at events – seemed like a good thing. Yet the example of the reusable drinks cup, designed to be distributed at large sporting and music events, can help us pinpoint how waste-pickers stand to lose out from such rental or rentier circular economy schemes. Coca-Cola is one of the worst serial offenders in terms of plastic pollution, being named in 2018 and 2019 as the most polluting brand in a global audit of plastic rubbish (Lerner 2019). Yet within its multinational operations one inevitably finds examples of good practice, such as a partnership with organised waste-pickers in Brazil during the 2014 FIFA World Cup. *Catador* cooperatives were contracted to collect recyclables, given a uniform, and paid a fee for the service (Globalrec 2014). Full bottles were sold by Coca-Cola and empty ones collected and sold by waste-pickers; a win-win situation that would have been impossible had returnable cups instead of non-returnable bottles been used. Indeed, the expansion of redeemable bottles by Coca-Cola Brazil puts a source of one of waste-pickers' principal raw materials at risk (Packaging Europe 2020).

In China, meanwhile, the concept of the circular economy (*xunhuan jingji*) has been enshrined in law since 2008 as part of a wider initiative to create an 'ecological civilisation' (*shentai wenming*). In Schultz and Lora-Wainwright's (2019) study of a new Circular Economy Park in Guiyu, smaller companies and workshops were replaced with larger enterprises either owned by the local elite or the state, with the authors concluding that the plant, through its reconfiguration of existing circuits, marginalised weaker actors and brought 'little in terms of additional circularity' (2019: 10). In his study of recycling in Beijing, Inverardi-Ferri (2018) argues that official government policy has effectively entailed the double dispossession of waste-pickers and artisanal recyclers. Pushed further and further out of the city by its rapid expansion and the construction of consecutive ring-roads, recyclers have also increasingly been denied access to the waste materials that constitute their livelihood. Whereas construction waste in Beijing effectively functioned as a commons in the first years of rapid economic growth, enabling the social mobility of impoverished rural migrants, the latter now find themselves dislocated from economic centres and dispossessed of resources that are channelled through formal circuits (Inverardi-Ferri 2018).

This is not solely a problem for waste-pickers in the Global South. Local authorities in the UK have made representations to the national government urging it not to allow companies such as Coca-Cola to establish their own closed circular economies for materials such as plastic bottles. PET is currently one of the few fractions of UK recyclate on which cash-strapped councils can make a decent return that they can then reinvest in their waste management costs. If companies are allowed to selectively take responsibility for their own forms of packaging, there is a real risk that they will cherry-pick the most valuable resources, leaving public authorities or waste-pickers with materials that have either zero or minimal market value. Where companies close the loop by retaining ownership over their products or packaging, there is no moment where such materials can either enter a waste commons or even become municipal property.

The pressure to close landfills that often forms part of circular economy policies arguably constitutes an even greater threat to waste-pickers. The European Commission's Circular Economy package, for example, has a binding target to landfill only 10 per cent of all waste by 2030 and a total ban on landfilling all separately collected waste. As we have seen, waste sent to dumps or landfill in the Global South is not necessarily wasted: it may be immediately 'unwasted' through the labour of waste-pickers. In Rio de Janeiro's Gramacho landfill, for example, informal waste-work was considered a 'stable refuge' by waste-pickers, 'constant, one of the most stable sources of income in their lives' (Millar 2015: 39). Gramacho's recent closure has resulted in the loss of this stability, compensated for by a one-off 'golden handshake' and jobs for a limited number of *catadores* in recycling plants (Passos Lima 2015).

The pressure to close landfills has opened up renewed sector opportunities for waste-to-energy technology companies and so-called chemical recycling or pyrolysis, where waste is transformed into either fuel or 'virgin' plastic, wax, and oils. Such methods enjoy qualified support from organisations like the EMF, which suggest that they might be a preferable alternative to landfill during a transition to greater recycling and re-use rates (World Economic Forum et al. 2016: 42). Multinational companies that control formerly public waste management systems in the Global North are aggressively promoting waste-to-energy technologies in the Global South as a greener alternative to landfill. In 2017, for instance, the French multinational Veolia, the largest waste management company in the world and the biggest operator in Britain, signed a contract to build and operate a major waste incinerator in Mexico City that was designed to supply energy to the twelve lines of the city's metro system. The deal was worth an estimated cumulative revenue of €886 million to Veolia over 30 years and guaranteed it a supply of two thirds of the mega-city's waste-stream. 'Waste can become a valuable resource', gushed the company's vice-president for Latin America as he described the plant as part of the struggle against climate change (Veolia 2017). The problem was that waste was already being recovered as a resource by the city's 10,000 waste-pickers, many of whom work at landfills recovering thousands of tonnes of material that is officially categorised as having been wasted (Espinosa Sánchez 2018). The Mexico City scheme was ultimately scrapped following a successful lobbying campaign that focused on the incinerator's high costs, the unconstitutionality of privatising waste management, and the damage that might be done to reduction and recycling strategies. Yet as Demaria and Todt (2020) set out, waste-to-energy technologies in the developing world still receive substantial public subsidies, such as those from the Kyoto Protocol Clean Development Mechanism (CDM). Africa's first incinerator was built in Ethiopia in 2018, leaving thousands of waste-pickers without a livelihood and campaigns against potential incinerators are currently being waged in Lebanon, Puerto Rico, Egypt, Kenya, Brazil, and India.

As Gutberlet and Carenzo (2020) suggest, waste-pickers in the Global South can be seen as already forming an important part of the circular economy given their role in diverting materials from landfill. It is also possible to include waste-pickers in explicit circular economy schemes, recognising their important role as 'circular economy pioneers' – this has occurred in limited ways in Santiago de Chile, Delhi and São Paulo (Noble 2019). Yet where waste-pickers are incorporated into formal waste management systems and obtain benefits in terms of health and safety, recognition, and even, sometimes, remuneration, the continuities with historic commons that I have highlighted here are often radically transformed or lost. Waste in such cases becomes a municipal or private property resource; *requeche* is replaced by wages; the vulnerability on which access to labour is based is formalised and codified; and refuges from wage labour are often enclosed.

New enclosures and commons sense

Many years ago, Eric Wolf argued that the capitalist system had first 'ransacked the world in its search for capital', then turned tribesmen and peasants into 'scavengers and beachcombers on the slag heaps of civilization' (1971: 3). We should be alert to the present danger that capital returns for another round of primitive accumulation, where rubbish has become resource and scavengers at the slag heap risk further dispossession. There is a burgeoning scholarship on the 'new enclosures' of the twenty-first century, such as that of intellectual property (May 2000) and carbon trading (Bond 2012: 689). The kind of enclosures that this book has detailed can be considered both old and new. Many involve the traditional practice of throwing up a fence around a piece of land and proclaiming it private property, denying the poor in the process the right to linger, dwell, inhabit, and common in the process. On the other hand, thinking of the waste container as a technology of enclosure, suggesting that the materials held inside might be worth enclosing, and proposing that excesses of production but also of behaviour are being enclosed in new recycling plants are more novel lines of enquiry.

One continuity between old and new enclosures that has emerged through this book's focus on waste is the control of diversity and hybridity. Lost with the enclosure of rural fields in England and Uruguay were diverse forms of land ownership and use, replaced by a standardised model of concentrated private ownership. At the same time, the diversity and uncontrolled hybridity of plant and animal species also diminished, replaced by the emergence of standardised agricultural commodities, such as the kilo of Uruguayan beef. The expectation of standardised, identical product lines is one of the central factors behind the creation of the colossal quantities of waste in modern capitalism (Blanchette 2015; Reno 2015).

The forms of enclosure described in Montevideo's waste management sector also aim to produce a homogeneous, uniformed waged worker, material that is either dumped as waste or sold as commodity, and a standardised model of waste collection and recycling. Enclosure thus moves against hybrid forms of waste management and material encounter, with informal clasificador practices deemed wasteful or inappropriate in order to justify dispossession (see Gidwani 2013; Gidwani and Reddy 2011). Rather than being unproductive per se, Montevidean waste-pickers are more often accused of engaging in forms of production that clash with ideas of infrastructural modernity. Contact with the unadulterated wastestream is deemed to be so undignified, and relations with informal sector intermediaries so exploitative, that dispossessing them and enclosing waste in sanitary landfills or recycling plants is justified. Even where the productive capacities of informal sector clasificadores are recognised, they are characterised as inefficient in comparison to the mechanised, collective, Taylorian plants into which some workers are transferred. Yet as we have seen, the productive superiority of the recycling plant over informal landfill labour is not in fact so clear-cut.

The increasing attention paid to wastes and waste management across the disciplines highlights that there is no single way of knowing or theorising waste. Yet such a conclusion can be depoliticising and disabling if it is not accompanied by an ethnographic exploration of the way that different actors know and represent waste within structures of power, and the political and economic consequences of such representations. This book has tried to get close to both the materiality of waste and the way that differently situated actors come to know it. A chain of associations that equates waste with risky matter, and discards with waste, leads to a paradoxical situation where the destruction of the use-value in things can be advocated and implemented as 'common-sense' environmental policy. Like the boy in the Emperor's New Clothes, clasificador recovery of *requeche* represents an empirical reality check, reminding us that many wasted things are in fact *sano*: fit for consumption and re-use. Beyond this, their good sense represents a popular and sensory 'way of knowing' that re-grounds epistemological debates about (waste) matter in the politics of the everyday.

In this vein, we might suggest that *requeche* consumption, as a type of popular knowledge, represents an example of Gramsci's (2005) *senso comune*. The sensory processes involved in determining what is good to eat would run contrary to common sense's Aristotelian origins, where this term named 'a supposed extra sense ... enabling us to organize the impressions received from the other five' (Crehan 2016: 43). Instead, the common sense of eating that which is *sano* emerges from the use of the tongue, eyes, nose, and fingers to determine whether something is safe for human consumption. Like Gramsci's *buen sentido*, a positive component embedded in *senso comune*, the knowledge involved in figuring this out, or indeed evaluating what metals can be extracted from an assemblage and how, are substantive forms of popular knowledge rather than the kind of cognitive structures to be found in Bourdieu's concept of habitus (Crehan 2016: 46). And per Gramsci's conceptualisation of *buen sentido*, clasificador practices, I would suggest, can be seen as forms of 'awareness born out of the concrete experience of subalternity ... the seeds from which new political narratives emerge' (Crehan 2016: 48–9).

If it seems far-fetched to suggest that a new politics can be generated out of the practice of eating rubbish, let us remember that brought together in the category of municipal solid waste are very often commodities stripped of their aura, de-fetishised objects (Kantaris 2016: 54) that according to some have to be re-fetishised in order to prevent us from seeing them for what they are: freely accessible and often perfectly usable goods that have been temporarily placed outside of the realms of commodity circulation (Barnard 2016). The liminal position of the ex- or not-quite commodity enables such instances of gift exchange, sharing, and ethical engagement as have been explored in this book, examples perhaps of the kind of relations that can be sustained if we are brave enough to tear down the fences of hygienic enclosure. The story is a hopeful one, because it suggests that rather than conceiving of the world as steadily and inexorably colonised by capital, we need but look around the back of the factory to discover a potential commons. *La plage* not so much *sous les pavés* as *dans les poubelles*.[1]

It is true that ex-commodities can very soon become re-commodified, whether converted to *requeche* at local flea markets or sold as *material* in national and international recycling chains. In considering waste as a commons, I have rejected David Harvey (2012) and others' conceptualisation of these as consisting of 'both collective and non-commodified' goods, in favour of the identification of a range of features drawn from the case of the English commons, including customary rights claimed by a vulnerable population, the refuge provided from wage labour, the importance of access/use over ownership/exchange, and the blurring of the lines between work and play. As E.P. Thompson (1991: 84) writes, English commoners were not 'primitive communists', and neither are Uruguayan *clasificadores* pure post-capitalist subjects. Nor, on the other hand, are they hyper-individualists living a dog-eat-dog existence around a hellish landfill. In fact, even the dogs abandoned at the landfill do not eat each other, forming instead a *canis familiaris* that lives from the bounty and care of the *madre cantera*.

My treatment of the commons at once territorialises, temporalises, and materialises the concept, bringing it down from the heady heights of certain strands of radical social theory to ground it in the muddy realities of landfill extraction. It territorialises, because it affirms, like Ostrum and her disciples (Ostrum 1991; Ostrum and Hess 2007; Ostrum et al. 1994) and like the English roots from which commons theory grows, that commons can be identified in territories and not only in the activity of 'commoning'. A feature of contemporary radical commons scholarship is the focus on the commons not as a resource but as an activity, a verb not a noun. For example, Harvey (2012: 73) argues that 'the common is not to be constructed … as a particular kind of thing, asset or even social process but as an unstable and malleable social relation'. The relation between clasificadores and waste is clearly important, but I resist the idea that, through their activity alone, clasificadores produce the waste commons, since the emphasis on commoning sits uneasily with the mutual dependence between clasificadores and the capitalist market in recyclable materials.

My framing also situates the commons within temporal processes by affirming that inclusion in the waste commons is but a temporary stage in the social life of a thing. It is the transient property status of waste, after decommodification but before re-appropriation, which allows clasificadores to claim a privileged place in a local moral economy as they compare themselves to shantytown neighbours who help themselves to materials that do have an owner. My framing also materialises the concept, grounding it in a debate over 'the material requirements for the construction of a commons-based economy' (Federici 2010: 4), and suggesting that commons can be embodied in the materials of the waste-stream. To a certain extent, my theorisation also provincialises current Western thinking that assumes commons to be something lost that must be recovered, as opposed to something actually existing that must be defended (see Berlant 2016). Throughout this book, I have attempted to draw in examples from beyond the pierced perimeter of the *cantera* in order to probe the utility of this framing of the commons and my conceptualisation of the waste commons beyond Uruguay. I would suggest that the concept of the commons put forward here enables a flexibility that allows for the identification of more commons, commoning, and commoners, while also preventing the moulding of such spaces and subjectivities so as to fit yearnings for purified post-capitalism over political and economic hybridity.

Given my focus on the customary claims to the waste commons made by clasificadores, it would make little sense to think of waste as a commons more generally without the presence of a group that struggles for access to the leftovers of productive processes. It is not just in the Global South that scavengers can be found, however, as evidenced by the burgeoning public focus and scholarship on freeganism, dumpster diving, skipping, and other forms of urban resource recovery in the Global North. I have argued that the distinction between those who scavenge or waste-pick by necessity and those who do so by choice should not be exaggerated. They are brought together in my very person, as I skipped bins in St Andrews and in Montevideo, forming a part of different communities in each instance and driven by different priorities but also by shared pleasures and moral logics of salvage and recuperation. They are also combined in my partner, an artist who for years scavenged for food and materials while she lived in an artist's squat in Cologne, and so had no problem doing so again when she moved to Montevideo, where she also found similarities between artist and clasificador discourses and practices about how discards enabled an autonomous lifestyle and a way to escape waged labour.

There are, then, clear parallels between the forms of recovery carried out by waste-pickers in the Global South and dumpster divers in the Global North. Edwards and Mercer (2013) note how the latter return to a use of their senses in the determination of whether food is safe to eat, something which, added to acceptable levels of food preparation experience, mean that very few fall sick. As we have seen, *gateadores* also enjoyed the 'element of surprise' and 'treasure hunt' aspects of recovering surplus (Edwards and Mercer 2013). Waste-pickers in Montevideo often have employment alternatives too – even if these are more restricted than those available to dumpster divers in the Global North – and the same has been argued for waste-pickers elsewhere (see Medina 2005; Millar 2015). Dumpster divers often face similar restrictions to accessing waste materials (O'Brien 2007) and they also frame their right to access in terms that would be familiar to my mates in the *cantera*: if someone has thrown something away and abandoned property rights over it, then why should I not be able to get some use or money out of it?

What if, alongside the kilo of quality beef, the clasificador way of life could also be considered a prime Uruguayan export destined for the Global North? What if a defence of waste as a source of livelihood, pride, and the good life could also be standardised in ways that protect access to this messy commons? What if horse-drawn transport and the recognition of urban natures were not relics of the past but signs from the future? What if, in the words of Padre Cacho, clasificadores were considered 'ecological prophets' (Alonso 1992)? Such a challenge would involve rethinking our relationship to waste, our ideas of a dignified life, our parameters of infrastructural modernity and contemporary economic relations. At a time when ecological collapse is forcing a radical rethinking of economic orthodoxy, it is crucially important that potentially fruitful concepts such as zero growth, zero waste, and the circular economy incorporate and build on the work of environmental pioneers like waste-pickers rather than constituting the basis of further enclosures.

One of the Intendencia de Montevideo's sanitary engineers once told me in an email exchange that he thought it was acceptable to 'work with rubbish' but not to 'work in rubbish'; an argument that he used to suggest that clasificadores should be able to work *with* rubbish in recycling plants but not *in* rubbish at the *cantera*. Throughout this book, we have encountered clasificadores who both live and work in an environment partly made up either of rubbish or of that which has at some point been classified as such. This includes houses partially composed of *requeche*, boots sunk into layers of tripe at the *cantera*, and homes surrounded by wasteland that doubles up as workplace. As such, we might say that clasificadores *inhabit* the waste that individuals, businesses, and the state shed in order to consume conveniently, make room for new stock, and maintain order in their own habitats.

As this book concludes, let me turn to a final concept, habitation, in order to further elucidate the processes of enclosure and dispossession that many waste-pickers and the waste commons are currently either undergoing or are at risk from. In *The Great Transformation*, Karl Polanyi (2001 [1944]: 36–7) quotes an English court document from 1607 that comments on the dilemma of socioeconomic change: 'The poor man shall be satisfied in his end: Habitation; and the gentleman not be hindered in his desire: Improvement.' For Polanyi, the formula 'hints at the tragic necessity by which the poor man clings to his hovel doomed by the rich man's desire for a public improvement which profits him privately' (2001 [1944]: 36–7). The context is Polanyi's discussion of the English enclosures that destroyed many a poor family's rooms as residents made way for the supposed improvement represented by more profitable animals.

The types of enclosure that I have discussed in this book have also often involved the enrichment of private interests like multinational corporations that profit from 'public improvements' to the city's waste management, financed in part so that urban elites might find their place in the shifting sands of 'infrastructural modernity'. Clasificadores, on the other hand, have had to wage multiple battles to defend the right to inhabit wastescapes and to incorporate waste into their habitations. Think of Sergio and his ragged, fluctuating band of dumped husbands, cats, dogs, and chickens that lived in Usina 5 until they were evicted. Or of Natalia Trastos, Nacho, La Negra, and the families of the Felipe Cardoso shantytown who had all built homes on squatted wasteland and, lacking legal tenancy rights, risked being evicted at any moment. Or indeed of Juan, Morocho, and others, who endured criticism of 'having a tip in their house' by including *requeche* furnishings or using their yards to classify *material*.

A desire for clasificadores to 'cling to their hovels' is a charge that could perhaps be levelled at me, given that I do not fully and uncritically embrace the current politico-religious orthodoxy regarding the dignifying of clasificador livelihoods. It is true that I find much in the current social life of Uruguayan waste worth salvaging. My interlocutors were, nevertheless, for the most part explicitly committed to ideas of progress, whether this was enabled through work in a recycling plant or the high earnings available through hard labour at the *cantera*. As Robert Marzec (2015: 85) notes, it is probably not true that English commoners rejected improvement either, and there is evidence that the pre-enclosure Saxon landholding system also led to ameliorations in soil quality and innovations in farming. So rather than inhabitancy defined as stagnation or 'clinging' to the status quo, I prefer Marzec's framing of it as 'lingering': 'the non-accumulative (non-profit driven) sustaining of life' and the 'sustainability of having access to a liveable occupation' (2015: 116). Is this not precisely the struggle of many clasificadores: to linger a little longer on the landfill or on city streets and to exercise an occupation that allows them to live?

Notes

Introduction: 'La Basura Es de los Pobres' – *'Rubbish Belongs to the Poor'*

1. Unión de Clasificadores de Residuos Urbanos Sólidos.
2. All names of persons in this book have been changed, as have those of a number of organisations and businesses. There is only one landfill in Montevideo, so I have decided not to change its name, nor that of its associated waste-picking cooperative, COFECA. A few nicknames have also remained unchanged.
3. *Mate* is a bitter tea, popular in Uruguay and Argentina, drunk from a gourd with a metal straw and often shared. In Uruguay, drinking *mate* is a veritable obsession, with people taking a flask and *mate* with them everywhere.

1 *'All Because We Bought Those Damn Trucks': Hygienic Enclosure and Infrastructural Modernity*

1. From 1908 until 2009, with some hiatuses, known as the Intendencia Municipal de Montevideo (IMM).
2. Decree. No.1585, Article 22e.
3. Ibid., Article 23.
4. Decree No. 11566.
5. Decree No.14.001(1967), Article 23.
6. One of the key theorists, Edwards, offers the disclaimer that he is principally writing about the 'developed world', not the Global South, where infrastructure might be much more precarious (2002: 188).
7. 'Our city is the only one that combines a large number of inhabitants with a dump just metres from its population, an unsanitary centre which directly and negatively affects hygiene and public health as well as revealing to tourists an immense and grave source of infectious disease germs, origin of the flies and rats which afflict the city ...' (Alfaro 1970a). Another intellectual, Hipólito Millot Grané, evoked a 'site of desolation and misery ... pestilence and infectious smoke ... clouds of flies and nauseous dust' (Alfaro 1971a).
8. See Alexander and Reno (2014) for a discussion of incineration and municipal socialist utopias in 1960s Sheffield.
9. They were *vagabundos* [vagabonds], *hurgadores* [rummagers], *cirujas, bichícomes*, or píchis. The last two terms reportedly stem from the English 'beachcomber' and might have originated in the beachcombers who searched El Cerro's beach for flotsam and jetsam that drifted in from the city's port. In any case, they are also words with unpleasant associations, homonymous with piss (*pichí*) and insects (*bichos*).

10. Figures are for 2013.
11. Uruguayans are particularly fond of using nicknames. This tendency is perhaps more pronounced in the world of clasificadores but I also observed an incident at the Laboratorio that indicates their more common usage. The director wanted to get in touch with a civil servant whom she only knew by the name of 'El Chino' (the Chinaman). Attempts to contact him using the general directory thus failed, and she had to phone around several colleagues to find out his real name.

2 *The Mother Dump: Montevideo's Landfill Commons*

1. There is no landfill tax in Uruguay. In the UK, in line with EU policy, such a tax is designed to discourage landfilling and promote alternatives, and currently sits at more than £90/ton.
2. *Pasta base* is a crude extract of the coca leaf which contains between 40 per cent and 91 per cent cocaine sulphate along with companion coca alkaloids. In recent years its consumption has become widespread among young men in low-income neighbourhoods of Montevideo (see Suárez et al. 2014).
3. Tom Neumark is a social anthropologist who has conducted research in a shantytown bordering Nairobi's Dandora landfill.

4 *Care, (Mis)Classification, and Containment at the Aries Recycling Plant*

1. The Frente Amplio has held the position ever since.
2. Instituto Nacional de Empleo y Formación Profesional.
3. Health insurance and pension contributions were 18.5 per cent, aguinaldo (an end-of-year Christmas bonus) was 8.33 per cent, holiday pay was 10.18 per cent.
4. IPUSA representatives told me that they had clients who sold them up to 500 tonnes of paper per month, whereas the plants sold them only 10 tonnes.
5. 'Open the books' – as in the historic trade union demand for employers to be transparent about a business's finances, in order to use this information to their advantage in collective bargaining.
6. The *pista* is the name given to the platform where trucks dump and clasificadores recover material at the landfill but Bolso apparently transfers the name to the plant in this quote.
7. Mbembe uses this term to denote colonial authority and 'the authoritarian modality par excellence' (1992: 3).

5 *Precarious Labour Organising and 'Urban* Alambramiento'

1. Some material from this chapter is reproduced with the kind permission of the publisher from the author's chapter: 'The Uruguayan recyclers' union: Clasificadores, circulation, and the challenge of mobile labor organizations', in *Urban Revolt: State Power and the Rise of People's Movements in the Global South*, edited

by Trevor Ngwane, Immanuel Ness and Luke Sinwell (New York: Haymarket, 2017).

2. Uruguay, in turn, is one of the few countries in the world to have every national trade union (one per sector) affiliated to a central national body. The federation had several unsuccessful predecessors, which splintered due to ideological differences (such as during the cold war) or repression (during the 1973–85 dictatorship). The federation in its current guise was founded after trade unions were legalised toward the end of the dictatorship. It held a victorious May Day Rally in 1984 under the banner of 'PIT-CNT: A Single Trade-Union Movement' (González Sierra 1989).
3. In some ways this can also be seen as a dispute between conflicting rights: the right of the clasificadores to work and circulate throughout the city versus animal rights and children's rights (some used a children's-rights-based discourse to criticise clasificadores who worked with their children and occasionally sent them inside containers to remove recyclables).
4. Lacalle Pou lost that election against Vásquez but defeated Frente Amplio candidate Daniel Martínez in 2020, breaking the Frente Amplio's 15-year grip on power.
5. As Wirth recounts, the term 'popular recycling' first appears in an MNCR document in 2009, and is defined as 'the construction of new models of "solidarity separation" and self-managed recycling that advance the role of workers in this productive chain at the same time as they serve as a reference for the construction of public policies that distribute the wealth originating from waste' (Wirth 2016: 149).

Conclusion: Circular Economies, New Enclosures, and the Commons

1. '*Sous les pavés, la plage* [under the paving stones, the beach]' was a situationist slogan from May '68, which made reference to both the sand that protesters encountered when pulling out paving stones, and their utopian illusions.

References

Note: *Many of the newspaper articles cited here were recovered from the library of the Junta Departamental de Montevideo. They had been cut out and page numbers had neither been conserved nor proved traceable.*

Abengoa. 2014. Sistemas de recolección de residuos diferenciados, viabilidad a escala local. Presentation given to CEMPRE (Compromiso Empresarial para el Medio Ambiente), https://docplayer.es/79423950-Teyma-medioambiente-sistemas-de-recoleccion-de-residuos-diferenciados-viabilidad-a-escala-local.html

Acción. 1968. Operativo limpieza: Un éxito. 18 October.

Ahora. 1972. ¿Qué pasa con la planta municipal de fertilizantes? 24 July.

Alexander, Catherine. 2009. Illusions of freedom: Polanyi and the third sector. In: Chris Hann and Keith Hart (eds) *Market and Society: The Great Transformation Today*. Cambridge: Cambridge University Press.

Alexander, Catherine and Joshua Reno (eds). 2012. *Economies of Recycling: Global Transformations of Materials, Values, and Social Relations*. London: Zed Books.

Alexander, Catherine and Josh Reno. 2014. From biopower to energopolitics in England's modern waste technology. *Anthropological Quarterly* 87(2): 335–358.

Alfaro, Hugo. 1971. Hurgando en el basural. *Marcha*, 17 September.

Alonso, Isidro. 1992. Profeta de la ciudad. *Novamerica* 53.

Alpini, Alfredo and Liliana Delfino. 2016. Higienismo en Montevideo (1829–1912): parques y espacios verdes (I). *Relaciones: Revista al Tema del Hombre* 383: 10–12.

Amin, Ash and Phillip Howell (eds). 2016. *Releasing the Commons: Rethinking the Futures of the Commons*. London: Routledge.

Anand, Nikhil, Akhil Gupta and Hannah Appel (eds). 2018. *The Promise of Infrastructure*. Durham, NC: Duke University Press.

Anderson, Benedict. 1983. *Imagined Communities*. London: Verso.

Appadurai, Arjun (ed.). 1986. *The Social Life of Things: Commodities in Cultural Perspective*. Cambridge: Cambridge University Press.

Arendt, Hannah. 1998 [1958]. *The Human Condition*. Chicago: University of Chicago Press.

Arneil, Barbara. 1994. Trade, plantations, and property: John Locke and the economic defense of colonialism. *Journal of the History of Ideas* 55(4): 591–609.

Astuti, Rita. 1998. 'It's a boy', 'It's a girl!' Reflections of sex and gender in Madagascar and beyond. In: Michael Lambek and Andrew Strathern (eds) *Bodies and Persons: Comparative Perspectives from Africa and Melanesia*. Cambridge: Cambridge University Press.

AUIP (Asociación Uruguaya de Industria del Plástico). 2020. Available at: www.auip.com.uy/historia-industria-del-plastico.php (accessed 1 April 2020).

Auyero, Javier and María Fernanda Berti. 2013. *La violencia en los márgenes: Una maestra y un sociólogo en el conurbano bonaerense*. Buenos Aires: Katz.

Auyero, Javier and Debora Swistun. 2007. Amidst garbage and poison: An essay on polluted peoples and places. *Contexts* 6(46): 46–51.

Bakhtin, Mikhail. 1984 [1965]. *Rabelais and His World*. Bloomington: Indiana University Press.

Barbero, Iker. 2015. When rights need to be (re)claimed: Austerity measures, neoliberal housing policies and anti-eviction activism in Spain. *Critical Social Policy* 35(2): 270–280.

Barles, Sabine. 2005. *L'Invention des déchets urbains: France: 1790–1970*. Paris: Champ Vallon.

Barnard, Alex V. 2016. *Freegans: Diving into the Wealth of Food Waste in America*. Minneapolis: University of Minnesota Press.

Barracchini, Hugo and Carlos Altezor. 2010. *Historia urbanística de la ciudad de Montevideo: Desde sus orígenes coloniales a nuestros días*. Montevideo: Trilce.

Barrán, José Pedro. 2014. *Historia de la sensibilidad en el Uruguay*. Montevideo: Ediciones de la Banda Oriental.

Bartesaghi, Ignácio and Susana Managa. 2012. China y Uruguay: Oportunidades y retos para vencer asimetrías. In: José Ignácio Martínez Cortés (ed.) *América Latina y el Caribe – China, relaciones políticas e internacionales*. Mexico City: Unión de Universidades de América Latina y el Caribe.

Bauman, Zygmunt. 2003. *Wasted Lives: Modernity and Its Outcasts*. London: Wiley.

Benjamin, Walter. 1999. *The Arcades Project*, trans. Howard Eiland and Kevin McLaughlin. Cambridge, MA: Harvard University Press.

Bennett, Jane. 2010. *Vibrant Matter: A Political Ecology of Things*. Durham, NC: Duke University Press.

Berlant, Lauren. 2016. The commons: Infrastructures for troubling times. *Environmental and Planning D* 34(3): 393–419.

Bértola, Luis. 2005. *A 50 años de la Curva de Kuznets: Crecimiento económico y distribución del ingreso en Uruguay y otros países de nuevo asentamiento desde 1870*. Instituto Aureliano Figuerola de Historia Económica, Working paper series No. 05-04, Universidad Carlos III de Madrid, Mayo 2005.

Bertoni, Reto. 2011. El modelo energético de la 'Suiza de América' como problema. Aportes de un análisis sectorial del consumo en Uruguay. *Revista Uruguaya de Historia Económica* 1(1): 76–102.

Birkbeck, Chris. 1978. Self-employed proletarians in an informal factory: The case of Cali's garbage dump. *World Development* 6(9–10): 1173–1185.

Blanchette, Alex. 2015. Herding species: Biosecurity, posthuman labor, and the American industrial pig. *Cultural Anthropology* 30(4): 640–669.

Boarder-Giles, David. 2014. The anatomy of a dumpster: Abject capital and the looking glass of value. *Social Text* 32(1): 93–113.

Boarder-Giles, David. 2015. The work of waste-making: Biopolitical labour and the myth of the global city. In: Jonathan Paul Marshall and Linda H. Connor (eds) *Environmental Change and the World's Futures: Ecologies, Ontologies and Mythologies*. London: Routledge.

Bond, Patrick. 2012. Emissions trading, new enclosures and eco-social contestation. *Antipode* 44(3): 684–701.

Bonino, Francisco A. 1958. *Eliminación de los residuos domiciliarios: Informe del Ingeniero Jefe de la Dirección de Limpieza y Usinas del Consejo Departamental de Montevideo*. Montevideo: Intendencia Municipal de Montevideo.

Bourgois, Phillipe. 1995. *In Search of Respect: Selling Crack in El Barrio*. Cambridge: Cambridge University Press.

Bowker, Geoffrey C. and Susan Leigh Star. 1999. *Sorting Things Out: Classification and Its Consequences*. Cambridge, MA: MIT Press.

BP. 1966. Montevideo ciudad antihigiénica. 24 January.

Breman, Jan. 2013a. A bogus concept? *New Left Review* 84: 128–130.

Breman, Jan. 2013b. *At Work in the Informal Economy of India*. Oxford: Oxford University Press.

Bryer, Alice. 2010. Beyond bureaucracies? The struggle for social responsibility in the Argentine workers' cooperatives. *Critique of Anthropology* 30(1): 41–61.

Bullard, Robert D. 1990. *Dumping in Dixie: Race, Class, and Environmental Quality*. Boulder, CO: Westview Press.

Butt, Waqas. 2019. Beyond the abject: Caste and the organization of work in Pakistan's waste economy. *International Labor and Working Class History* 95: 18–33.

Calvino, Italo. 2009. *The Road to San Giovanni*. London: Penguin.

Cambadu. 2016. Proyecto oblige supermercados a donar comida a punto de vencer. Available at: www.cambadu.com.uy/index.php/2016/07/04/supermercados-desperdicios-comida/ (accessed 2 April 2020).

Cant, Alana. 2019. *The Value of Aesthetics: Oaxacan Woodcarvers in Global Economies of Culture*. Austin: University of Texas Press.

Carenzo, Sebastián. 2016. Waste classification as a craft under construction: The worker's experience at Buenos Aires' 'Social Classification Plants'. *Journal of Latin American and Caribbean Anthropology* 21(2): 276–293.

Carenzo, Sebastián and Pablo Miguéz. 2010. De la atomización al asociativismo: Reflexiones en torno a los sentidos de la autogestión en experiencias asociativas desarrolladas por cartoneros/as. *Maguare* 24: 233–263.

Carrasco, Sansón, 2006 [1883]. La basura. In: Sansón Carrasco and Claudio Paolini. *Crónicas de un fin de siglo por el montevideano Sansón Carrasco (1892–1909)*. Montevideo: Ediciones de la Banda Oriental. Originally published in *La Razón de Montevideo*, 1 August 1883.

Carsten, Janet. 1995. The substance of kinship and the heat of the hearth: Feeding, personhood, and relatedness among the Malays in Pulau Langkawi. *American Ethnologist* 22(2): 223–241.

Carsten, Janet. 2013. What kinship does – and how. *Hau: Journal of Ethnographic Theory* 3(2): 245–251.

Centner, Ryan. 2012. Techniques of absence in participatory budgeting: Space, difference and governmentality across Buenos Aires. *Bulletin of Latin American Research* 31(2): 142–159.

Chabalgoity, Manuel et al. 2004. Gestión de Residuos Sólidos Urbanos, un abordaje territorial desde la perspectiva de la inclusión social, el trabajo y la producción, *PAMPA* 1(2): 37–84.

Chalfin, Brenda. 2014. Public things, excremental politics, and the infrastructure of bare life in Ghana's city of Tema. *American Ethnologist* 41(1): 92–109.

Chen, Martha Alter. 2012. The informal economy in comparative perspective. In: James G. Carrier (ed.) *A Handbook of Economic Anthropology*, 2nd edn. Cheltenham: Edward Elgar.

Chibnik, Michael. 2011. *Anthropology, Economics and Choice*. Austin: University of Texas Press.

Choudary, Bikramaditya Kumar. 2003. Waste and waste-pickers. *Economic and Political Weekly* 38(5): 5240–5242.

Clara, Mercedes. 2012. *Padre Cacho: Cuando el otro quema adentro*. Montevideo: Trilce/ OBSUR.

Clark, J.F.M. 2007. 'The incineration of refuse is beautiful': Torquay and the introduction of municipal refuse destructors. *Urban History* 34: 255–277.

Classen, Constance, David Howes and Anthony Synnott (eds). 1994. *Aroma: The Cultural History of Smell*. London: Routledge.

Clifford, James. 1989. The others: Beyond the 'salvage' paradigm. *Third Text*, 3(6): 73–78.

Collier, Stephen J. 2011. *Post-Soviet Social: Neoliberalism, Social Modernity, Biopolitics*. Princeton, NJ: Princeton University Press.

Collins, Thomas. 1988. An analysis of the Memphis garbage strike of 1968. In: Johnetta B. Cole (ed.) *Anthropology for the Nineties*. New York: Free Press.

Connell, R.W. 1995. *Masculinities*. Cambridge: Polity Press.

Cooper, Ben. 2008. The Molendinar Burn. Available at: http://catchingphotons.co.uk/blog/miscellaneous/the-molendinar-burn/ (accessed 20 August 2021).

Corwin, Julia. 2019. Between toxics and gold: Devaluing informal labor in the global urban mine. *Capitalism Nature Socialism* 31(4): 106–123.

Crehan, Kate. 2016. *Gramsci's Common Sense: Inequality and Its Narratives*. Durham, NC: Duke University Press.

Dagognet, Francois. 1997. *Des detritus, des déchets, de l'abject: Une philosophie écologique*. Paris: Le Plessis-Robinson.

Dalakoglou, Dimitris. 2016. Infrastructural gap. *City* 20(6): 822–831.

Dalakoglou, Dimitris and Yannis Kallianos. 2014. Infrastructural flows, interruptions and stasis in Athens of the crisis. *City* 18(4–5): 526–532.

Darlington, Ralph. 2013. The role of trade unions in building resistance: Theoretical, historical, and comparative perspectives. In: Maurizio Atzeni (ed.) *Workers and Labour in a Globalized Capitalism*. London: Red Globe Press.

Davies, Steve. 2007. *Politics and Markets: The Case of UK Municipal Waste Management*, School of Social Sciences Working Paper 95. Cardiff: University of Cardiff.

DEFRA (Department for Environment, Food and Rural Affairs). 2012. Guidance on the legal definition of waste and its application. Available at: www.gov.uk/topic/environmental-management/waste (accessed 23 April 2019).

Deleuze, Gilles and Claire Parnet. 1987. *Dialogues*. London: Athlone Press.

de los Santos, Federico. 2016. El almuerzo desnudo. *La Diaria*, 7 July. Available at: https://ladiaria.com.uy/articulo/2016/7/el-almuerzo-desnudo/ (accessed 1 April 2020).

Demaria, Federico and Marcos Todt. 2020. How waste pickers in the global south are being sidelined by new policies. *The Conversation*. Available at:

https://theconversation.com/how-waste-pickers-in-the-global-south-are-being-sidelined-by-new-policies-132521 (accessed 1 April 2020).

Demos, T.J. 2013. *Return to the Postcolony: Spectres of Colonialism in Contemporary Art*. Berlin/ New York: Sternberg Press.

de Soto, Hernando. 2002. *The Other Path: The Economic Answer to Terrorism*. New York: Basic Books.

Devine, T.M. 2018. *The Scottish Clearances: A History of the Dispossessed*. London: Allen Lane.

Dickens, Charles. 1997 [1865]. *Our Mutual Friend*. London: Penguin.

Dimarco, Sabina. 2011. Entre riesgo social y beneficio ambiental: Transformaciones sociohistóricas en la construcción social del riesgo en la clasificación de residuos. *Quid* 16(2): 162–180.

Dinler, Demet S. 2016. New forms of wage labour and struggle in the informal sector: The case of waste pickers in Turkey. *Third World Quarterly* 37(10): 1834–1854.

Dirección Nacional de Evaluación y Monitoreo. 2006. *Perfil social de clasificadores inscriptos en el PANES Programa 'Uruguay Clasifica'*. Documento de Trabajo IV – Serie Perfiles 4 (Caracterización de la Población). Montevideo, Uruguay.

Di Stefano, Eugenio. 2012. From shopping malls to memory museums: Reconciling the recent past in the Uruguayan neoliberal state. *Dissidences: Hispanic Journal of Theory and Criticism* 4(8).

Douglas, Mary. 1992. *Risk and Blame: Essays in Cultural Theory*. London: Routledge.

Douglas, Mary. 2002 [1966]. *Purity and Danger*. London: Routledge.

Downey, Greg, Monica Dalidowicz and Paul H. Mason. 2014. Apprenticeship as method: Embodied learning in ethnographic practice. *Qualitative Research* 15(2): 183–200.

Drotbohm, Heike and Erdmute Alber. 2015. Introduction. In: Erdmute Alber (ed.) *Anthropological Perspectives on Care: Work, Kinship and the Life-course*. London: Palgrave Macmillan.

Dumont, Louis. 1986. *Essays on Individualism: Modern Ideology in Anthropological Perspective*. Chicago: University of Chicago Press.

Duviols, Jean-Paul. 1985. *L'Amérique espagnole vue et revée: Les livres de voyages de Christophe Colomb à Bougainville*. Paris: Editions Promodis.

EC DG Environment (European Commission Director General Environment). 2012. Guidance on the interpretation of key provisions of Directive 2008/98/EC in waste. Available at: https://ec.europa.eu/environment/waste/framework/guidance.htm (accessed 1 April 2020).

Edgerton, David. 2008. *The Shock of the Old: Technology and Global History Since 1900*. London: Profile Books.

Edgerton, David. 2018. Turning the global history of technology upside down: The supremacy of Uruguay. Talk given at the Centre for Global Knowledge Studies (Gloknos), 19 October Available at: www.crassh.cam.ac.uk/gallery/audio/david-edgerton-turning-the-global-history-of-technology-upside-down (accessed 20 August 2021).

Edwards, Ferne and Dave Mercer. 2013. Food waste in Australia: The freegan response. In: David Evans, Hugh Campbell and Anne Murcott (eds) *Waste Matters: New Perspectives on Food and Society*. London: Wiley-Blackwell.

Edwards, Paul N. 2002. Infrastructure and modernity: Force, time and social organization in the history of sociotechnical systems. In: Thomas J. Misa, Phillip Brey and Andrew Feenburg (eds) *Modernity and Technology*. Cambridge, MA: MIT Press.

El Debate. 1968. Protestan contra una medida del municipio. 22 March.

El Día. 1970. Basural gigantesco en Carrasco. 19 November.

El Día. 1973a. Limpieza de la ciudad (III): ¿Es realmente un 'fertilizante' el producto transformado de la basura? 3 April.

El Día. 1973b. Limpieza de la ciudad (IV): Inconvenientes del producto que resulta de los residuos tratados. Lo que los críticos maliciosos no dicen de la lucha contra el basural. 6 April.

El Día. 1976. IMM: A los hurgadores de residuos domiciliarios. 15 June.

El Día. 1980. No, no puede ser verdad. 22 August.

El Diario. 1968. Basura: Convertiran 400 tons. por día en fertilizante. 19 October.

El Diario. 1969. Dinamitan chimeneas e intentan volarlas. 8 April.

El Diario. 1972. Fertilizantes: Produciran 20 millones al municipio. 13 July.

El Diario. 1974. La intendencia inicia 'la batalla del plástico' para poder aprovechar la basura. 28 April.

El Diario. 1979. Municipio vs. Junta-papeles. 5 April.

El Observador. 2014a. Clasificadores ganarán $18.000 en nuevas plantas de reciclaje. 24 March. Available at: www.elobservador.com.uy/clasificadores-ganaran-18000-nuevas-plantas-reciclaje-n274682 (accessed 1 April 2020).

El Observador. 2014b. Basura en Montevideo: el antes y el después. 6 May. Available at: www.elobservador.com.uy/basura-montevideo-el-antes-y-el-despues-n277867 (accessed 1 April 2020).

El País. 1967. Batalla de la mugre (II) crematorios, canteras o fertilizantes: ¿Qué haremos ahora con nuestra basura? 28 April.

El País. 1971. Montevideo es invadida otra vez por la basura. 25 July.

El País. 1980. Se está reprimiendo la acción de los hurgadores de residuos. 8 August.

El País. 1989. Intendencia responsable del 90 per cent de la recolección en Montevideo. 24 July.

El País. 2003. Contenedores invadarán Montevideo. 9 July.

El País. 2014. Gobierno ultima licitación de gestión de residuos y generación eléctrica. 21 April.

El País. 2016. Montevideo, qué mal te veo. 6 June. Available at: www.elpais.com.uy/opinion/editorial/montevideo-que-mal-te-veo.html (accessed 1 April 2020).

El Popular. 1972. Fertilizantes: Iniciativa del Frente Amplio empieza a dar sus frutos. 10 July.

EMF (Ellen MacArthur Foundation). 2020. Concept. Available at: www.ellenmacarthurfoundation.org/circular-economy/concept (accessed 1 April 2020).

Epoca. 1966. Montevideo: ¿Suiza o sucia de América? 5 May.

Errandonea, Alfredo and Daniel Costabile. 1968. *Sindicato y sociedad en el Uruguay*. Montevideo: Biblioteca de Cultura Universitaria.

Espinosa Sánchez, Tania. 2018.Threat of new waste incinerator in Mexico City puts informal waste pickers' livelihoods at risk. Available at: www.wiego.org/blog/threat-new-waste-incinerator-mexico-city-puts-informal-waste-pickers-livelihoods-risk (accessed 1 April 2020).

European Commission (2017) Notice. EU Guidelines on food donation (2017/C 361/01). *Official Journal of the European Union* 60(C361): 1–30.

Evans, David, Hugh Campbell and Anne Murcott. 2013. A brief pre-history of food waste and the social sciences. In: David Evans, Hugh Campbell and Anne Murcott (eds) *Waste Matters: New Perspectives on Food and Society*. London: Wiley-Blackwell.

Faraone, Roque. 1986 [1974]. *De la prosperidad a la ruina: Introducción a la historia económica del Uruguay*. Montevideo: Arca.

Federici, Silvia. 2010. Feminism and the politics of the commons. *The Commoner*. Available at: http://wealthofthecommons.org/essay/feminism-and-politics-commons (accessed 1 April 2020).

Fernández y Medina, Benjamin. 1904. *Ley organica de las juntas economicas adminstrativas*, Tomo 1. Montevideo: A. Barreiro y Ramos.

Fernández, Lucía. 2007. De hurgadores a clasificadores organizados: Análisis político institucional del trabajo con la basura en Montevideo. In: Pablo J. Schamber and Francisco M. Suárez (eds) *Recicloscopio*. Buenos Aires: Promoteo.

Fernández, Lucía. 2010. Dynamiques du recyclage spontané: Regards croisés sur les villes de Montevideo et Paris au XIXe. Master's thesis, École Nationale Supérieure de'Architecture de Grenoble.

Fernández, Lucía. 2012. *Paisajes-basura: Dinámicas y externalidades territoriales del reciclaje en Montevideo Uruguay*. Documento de Trabajo de WIEGO (Políticas Urbanas) No. 25. Manchester: WIEGO.

Filgueria, Fernando. 1995. *A Century of Social Welfare in Uruguay: Growth to the Limit of the Batllista Social State*, trans. Judy Lawton. Kellog Institute Democracy and Social Policy Series. Notre Dame: University of Notre Dame Press.

Fonseca, Claudia. 2000. *Familia, Fofoca y honra*. Porto Alegre: Editora da Universidade Federal da Rio Grande do Sul.

Foucault, Michel. 2007. *Security, Territory, Population: Lectures at the Collège de France, 1977–78*. Basingstoke: Palgrave Macmillan.

Fraser, Nancy. 2016. Contradictions of capital and care. *New Left Review* 100: 99–119.

Fredericks, Rosalind. 2014. Vital infrastructures of trash in Dakar. *Comparative Studies of South Asia, Africa and the Middle East* 34(3): 532–548.

Frégier, Honoré. 1840. *Des classes dangereuses de la population dans les grandes villes*. Paris: Baillière.

Furedy, Christine. 1990. Social aspects of solid waste recovery in Asian cities. *Environmental Sanitation Review* 30: 2–52.

Furniss, Jamie. 2017. What type of problem is waste in Egypt? *Social Anthropology* 25(3): 301–317.

Gago, Veronica. 2017. *Neoliberalism from Below: Popular Pragmatics and Baroque Economies*. Durham, NC: Duke University Press.

Gall, Gregor (ed.). 2002. *Union Organizing: Campaigning for Trade Union Recognition*. London: Routledge.

Gandy, Matthew. 2003. *Concrete and Clay: Reworking Nature in New York City*. Cambridge, MA: MIT Press.

Gidwani, Vinay. 2013. Six theses on waste, value, and commons. *Social and Cultural Geography* 14(7): 773–783.

Gidwani, Vinay. 2015. The work of waste: Inside India's infra-economy. *Transactions of the Institute of British Geographers* 40: 575–595.
Gidwani, Vinay and Rajyashree N. Reddy. 2011. The afterlives of 'waste': Notes from India for a minor history of capitalist surplus. *Antipode* 45(5): 1625–1658.
Gille, Zsuzsa. 2007. *From the Cult of Waste to the Trash Heap of History: The Politics of Waste in Socialist and Postsocialist Hungary*. Indianopolis: Indiana University Press.
Gille, Zsuzsa 2010. Actor networks, modes of production, and waste regimes: reassembling the macro-social. *Environment and Planning A* 42: 1049–1064.
Gille, Zsuzsa. 2013. From risk to waste. In: David Evans, Hugh Campbell and Anne Murcott (eds) *Waste Matters: New Perspectives on Food and Society*. London: Wiley-Blackwell.
Gille, Zsuzsa. 2015. Ecological modernization of waste-dependent development? Hungary's 2010 red mud disaster. In: Ruth Oldenziel and Helmuth Trischler (eds) *Cycling and Recycling: Histories of Sustainable Practices*. London: Berghahn.
Globalrec. 2014. 840 waste pickers contracted to recycle during the World Cup. Available at: https://globalrec.org/2014/06/11/840-waste-pickers-contracted-to-recycle-during-the-world-cup/ (accessed 1 April 2020).
Gmelch, Sharon. 1986. Groups that don't want in: Gypsies and other artisan, trader, and entertainer minorities. *Annual Review of Anthropology* 15: 307–330.
Goldstein, Jesse. 2013. Terra economica: Waste and the production of enclosed nature. *Antipode* 45(2): 357–337.
González, Mike. 2018. *The Ebb of the Pink Tide: The Decline of the Left in Latin America*. London: Pluto Press.
González Sierra, Yamandú. 1989. *Reseña histórica del movimiento sindical uruguayo (1870–1984)*. Montevideo: CIEDUR.
Gowan, Teresa. 2010. *Hobos, Hustlers and Backsliders: Homeless in San Francisco*. Minneapolis: University of Minnesota Press.
Graeber, David. 2001. *Toward an Anthropological Theory of Value: The False Coin of our Own Dreams*. New York: Palgrave.
Graham, Steve and Simon Marvin. 2001. *Splintering Urbanism: Networked Infrastructures, Technological Mobilities and the Urban Condition*. London: Routledge.
Gramsci, Antonio. 2005. *Selections from the Prison Notebooks*. London: Lawrence and Wishart.
Gray, Alasdair. 2011 [1981]. *Lanark: A Life in 4 Books*. Edinburgh: Canongate.
Grimson, Alejandro. 2008. The making of new urban borders: Neoliberalism and protest in Buenos Aires. *Antipode* 40(4): 504–512.
Grisaffi, Tom. 2019. *Coca Yes, Cocaine No*. Durham, NC: Duke University Press.
Gudynas, Eduardo. 2004. Regresaron los soñadores de chimeneas. *Boletín en Ecología Social y Ecología Humana* 38.
Guha-Khasnobis, Basudeb, Ravi Kanbur and Elinor Ostrom. 2006. Introduction. In: *Linking the Formal and Informal Economy: Concepts and Policies*. Oxford: Oxford University Press.
Gutberlet, Jutta and Sebastián Carenzo. 2020. Waste pickers at the heart of the circular economy: A perspective of inclusive recycling from the Global South. *Worldwide Waste: Journal of Interdisciplinary Studies* 3(1): 1–14.

Hammond, J.L. and Barbara Hammond. 1987 [1911]. *The Village Labourer, 1760–1832*. London: Sutton Publishing.

Han, Clara. 2012. *Life in Debt: Times of Care and Violence in Neoliberal Chile*. Berkeley: University of California Press.

Hann, Chris and Keith Hart (eds). 2009. *Market and Society: The Great Transformation Today*. Cambridge: Cambridge University Press.

Hansen, Karen Tranberg, Walter E. Little and B. Lynne Milgram. 2013. *Street Economies in the Urban Global South*. Santa Fe: SAR Press.

Hardin, Garrett. 1968. The tragedy of the commons. *Science* 162(3859): 1243–1248.

Harris, Mark (ed.). 2007. *Ways of Knowing: New Approaches in the Anthropology of Knowledge and Learning*. London: Berghahn.

Hart, Keith. 1973. Informal income opportunities and urban employment in Ghana. *Journal of Modern African Studies* 11(1): 61–89.

Hart, Keith. 2008. Informal economy. In: Steven N. Durlauf and Lawrence E. Blume. *The New Palgrave Dictionary of Economics*. London: Palgrave Macmillan.

Hartmann, Chris. 2018. Waste picker livelihoods and inclusive neoliberal municipal solid waste management policies: The case of the La Chureca garbage dump site in Managua, Nicaragua. *Waste Management* 71: 565–577.

Harvey, David. 2008. The right to the city. *New Left Review* 53: 23–40.

Harvey, David. 2011. The future of the commons. *Radical History Review* 109: 101–107.

Harvey, David. 2012. *Rebel Cities*. London: Verso.

Harvey, Penelope. 2013. The material politics of solid waste. In: Penny Harvey (ed.) *Objects and Materials*. London: Routledge.

Harvey, Penelope. 2017. Waste futures: Infrastructures and political experimentation in southern Peru. *Ethnos* 82(4): 672–689.

Hawkins, Gay. 2003. Down the drain: Shit and the politics of disturbance. In: Gay Hawkins and Stephen Muecke (eds) *Culture and Waste: The Creation and Destruction of Value*. Oxford: Rowman and Littlefield.

Hawkins, Gay. 2006. *The Ethics of Waste*. Lanham, MD: Rowman and Littlefield .

Hechos. 1965. El Municipio rellena canteras con basura. 6 March.

Hecht, Gabrielle. 2012. *Being Nuclear: Africans and the Global Uranium Trade*. Cambridge, MA: MIT Press.

Hernandez, Vladimir. 2012. José Mujica: The world's 'poorest' president. *BBC Mundo*, 12 November. Available at: www.bbc.co.uk/news/magazine-20243493 (accessed 1 April 2020).

Hertzfeld, Michael. 2010. Engagement, gentrification, and the neoliberal hijacking of history. *Current Anthropology* 51(2): S259–S267.

Heynen, Nik, Maria Kaika and Erik Swyngedouw. 2006. *In the Nature of Cities: Urban Ecology and the Politics of Urban Metabolism*. New York: Routledge.

Hicks, Bill. 1997. 'Hooligans', track on: Bill Hicks, *Arizona Bay*. Salem: Rykodisc.

Hird, Myra J. 2012. Knowing waste: Towards an inhuman epistemology. *Social Epistemology* 26(3–4): 453–469.

Hird, Myra J. 2017. Waste, environmental politics and dis/engaged publics. *Theory, Culture & Society* 34(2–3): 187–209.

Horne, R.H. 1850. Dust; or, Ugliness redeemed. *Household Words*, 13 July.

Howarth, Anthony Leroyd. 2018. *A Travellers' Sense of Place in the City*. PhD thesis, University of Cambridge.

Howell, Signe. 2003. Kinning: The creation of life trajectories in transnational adoptive families. *Journal of the Royal Anthropological Institute* 9(3): 465–484.

ILO (International Labour Organization). 2013. *El desarrollo sostenible, el trabajo decent y los empleos verdes*. Geneva: International Labour Organization.

IM (Intendencia de Montevideo). 2015. *Trayectos Montevideanos: Inclusión social de hombres y mujeres clasificadores 2010–2015*. Montevideo: Intendencia de Montevideo.

IM (Intendencia de Montevideo). 2020. Residuos de empresas, comercios y organizaciones. Available at: https://montevideo.gub.uy/areas-tematicas/gestion-de-residuos/residuos-de-empresas-comercios-y-organizaciones (accessed 13 April 2020).

IMM/PNUD. 1996. *Clasificación y reciclo de residuos sólidos, Tomo I y II*. Montevideo: Coordinador Equipo Consultor.

IM (Intendencia de Montevideo), PNUD and PNUMA. 2012. *Caracterización de la población de clasificadores de residuos de Montevideo*. Montevideo: Intendencia de Montevideo.

Ingold, Tim. 1993. The art of translation in a continuous world. In: Gísli Pálsson (ed.) *Beyond Boundaries*. Oxford: Berg.

Ingold, Tim. 2000. *The Perception of the Environment: Essays on Livelihood, Dwelling and Skill*. London: Routledge.

Ingold, Tim. 2011. Reply to David Howes. *Social Anthropology/Anthropologie Sociale* 19(3): 323–327.

Inverardi-Ferri, C. 2017. The enclosure of 'waste land': Rethinking informality and dispossession. *Transactions of the Institute of British Geographers* 43: 230–244.

IP and MA. 2012. Caracterización de la población de clasificadores de residuos de Montevideo. Available at: https://montevideo.gub.uy/sites/default/files/caracterizacion_de_la_poblacion_de_clasificadores_de_residuos_de_montevideo.pdf (accessed 7 September 2021).

James, Deborah. 2014. *Money from Nothing: Indebtedness and Aspiration in South Africa*. Stanford, CA: Stanford University Press.

Johnson, Craig. 2004. Uncommon ground: The 'poverty of history' in common property discourse. *Development and Change* 35(3): 407–433.

Kalb, Don. 2017. Afterword: After the commons – commoning! *Focaal – Journal of Global and Historical Anthropology* 79: 67–73.

Kantaris, Geoffrey. 2016. Waste not, want not: Garbage and the philosopher of the dump (Wasteland and Estamira). In: Christoph Lindner and Miriam Meissner (eds) *Global Garbage: Urban Imaginaries of Waste, Excess, and Abandonment*. London: Routledge.

Kar, Sohini. 2018. *Financializing Poverty: Labor and Risk in Indian Microfinance*. Stanford, CA: Stanford University Press.

Kasmir, Sharryn and August Carbonella. 2008. Dispossession and the anthropology of labor. *Critique of Anthropology* 28(1): 5–25.

Katz, Cindy. 2004. *Growing Up Global*. Minneapolis: University of Minnesota Press.

Kelly, John and Edmund Heery.1989. Full-time officers and trade union recruitment. *British Journal of Industrial Relations* 27(2): 196–213.

Keskula, Eeva. 2014. Disembedding the company from kinship: Unethical families and atomised labour in an Estonian mine. *Laboratorium: Russian Review of Social Research*, 6(2): 58–76.

Lakoff, George. 1987. *Women, Fire, and Dangerous Things: What Categories Reveal about the Mind*. Chicago: University of Chicago Press.

La Mañana. 1971. La Usina No. 5 de limpieza tiene dificultades para funcionar. 12 January.

La Mañana. 1985. Dr. Lanza: El problema de los residuos es el más importante. 15 October.

Lambek, Michael. 2011. Kinship as gift and theft: Acts of succession in Mayote and Ancient Israel. *American Ethnologist* 38(1): 2–16.

Lankford, Bruce. 2013. *Resource Efficiency Complexity and the Commons: The Paracommons and Paradoxes of Natural Resource Losses, Wastes and Wastages*. Abingdon: Earthscan Publications.

La Plata. 1966. Deficiente higiene de la ciudad. 6 May.

La Red 21. 2007. El 'enterrado' de La Tablada. 16 July. Available at: www.lr21.com.uy/politica/265671-el-enterradero-de-la-tablada (accessed 1 April 2020).

Larkin, Brian. 2013. The politics and poetics of infrastructure. *Annual Review of Anthropology* 42: 327–343.

Latour, Bruno. 2004. *Politics of Nature: How to Bring the Sciences into Democracy*, trans. Catharine Porter. Cambridge, MA: Harvard University Press.

Lazar, Sian. 2008. *El Alto, Rebel City: Self and Citizenship in Andean Bolivia*. Durham, NC: Duke University Press.

Lazar, Sian. 2012. Disjunctive comparison: Citizenship and trade unionism in Bolivia and Argentina. *Journal of the Royal Anthropological Institute* 18: 349–368.

Le Bas, Damien. 2014. New scrap metal law to hit travellers hard. *Travellers Times*. Available at: www.travellerstimes.org.uk/news/2014/11/new-scrap-metal-law-hit-travellers-hard (accessed 16 April 2019).

Lefebvre, Henri. 1968. *Le Droit a la ville*. Paris: Seuil.

Lema, Patricia et al. 2017. *Informe final: Estimación de pérdidas y desperdicio de alimentos en el Uruguay: alcance y causas*. Programa Estratégico 4: Output 40202. Montevideo: Organización de las Naciones Unidas para la Agricultura y la Alimentación (FAO).

Lerner, Steve. 2010. *Sacrifice Zones: The Front lines of Toxic Chemical Exposure in the United States*. Cambridge, MA: MIT Press.

Lerner, Sharon. 2019. Coca-cola named most polluting brand in global audit of plastic waste. *The Intercept*. Available at: https://theintercept.com/2019/10/23/coca-cola-plastic-waste-pollution/ (accessed 10 April 2020).

Levine, Peter. 2002. Building the electronic commons. *The Good Society* 11(3): 4–9.

Lewis, Oscar. 1966. The culture of poverty. *Scientific American* 215(4): 19–25.

Liboiron, Max. 2019. Waste is not 'matter out of place'. Available at: https://discardstudies.com/2019/09/09/waste-is-not-matter-out-of-place/ (accessed 17 September 2020).

Linebaugh, Peter. 2003 [1991]. *The London Hanged*. London: Verso.

Linebaugh, Peter. 2008. *The Magna Carta Manifesto: Liberties and Commons for All*. Berkeley: University of California Press.

Linebaugh, Peter. 2014. *Stop Thief! The Commons, Enclosures and Resistance.* Oakland: PM Press.

Linebaugh, Peter. 2019. *Red Round Globe Hot Burning: A Tale at the Crossroads of Commons and Enclosure, of Love and Terror, of Race and Class, and of Kate and Ned Despard.* Oakland, CA: University of California Press.

LKSur Asociados, Alcoholes del Uruguay, and the Dirección Nacional de Energia. 2013. *Estudio de caracterización de residuos sólidos urbanos con fines energéticos.* Informe 1, Febrero 2013. Especificación Técnica No. 12047-ET-01. Montevideo: LKSur.

Locke, John. 1993 [c. 1681]. *Political Writings*, ed. David Wooton. Indianapolis: Hackett Publishing.

Locke, John. 2005 [1689]. *Two Treatises of Government & A Letter Concerning Toleration.* Stilwell: Digireads.

Lombardi, María José. 2006. El reciclador marginado: Un análisis sobre la percepción de los residuos y los clasificadores informales. *Anuario de Antropología Social y Cultural en el Uruguay* 5. Montevideo: Nordan Comunidad.

López Reilly, Andres. 2017. IMM impide con policías ingreso de hurgadores a Felipe Cardoso. *El País*, 14 February. Available at: www.elpais.com.uy/informacion/imm-impide-policias-ingreso-hurgadores.html (accessed 16 March 2017).

López Reilly, Andres. 2018. IMM pagará US$12 millones a empresa por limpieza céntrica. *El País*, 9 August. Available at: www.elpais.com.uy/informacion/sociedad/imm-pagara-us-millones-empresa-limpieza-centrica.html (accessed 2 April 2020).

Macedo, Joseli. 2012. Introduction: The urban divide in Latin America: Challenges and strategies for social inclusion. *Bulletin of Latin American Research* 31(2): 139–141.

MacBride, Samantha. 2011. *Recycling Reconsidered: The Present Failure and Future Promise of Environmental Action in the United States.* Cambridge, MA: MIT Press.

McCay, Bonnie J. and James M. Acheson (eds). 1987. *The Question of the Commons: The Culture and Ecology of Communal Resources.* Tucson: University of Arizona Press.

McKinnon, Catherine and Fenella Cannell (eds). 2013. *Vital Relations: Modernity and the Persistent Life of Kinship.* Santa Fe, NM: School for Advanced Research Press.

Magalhaes, Beatriz Judice. 2016. Liminaridade e exclusao: Caracterizaco permanente ou transitoria das relacoes entre os catadores e a sociedade brasileira? In: Jacquetto Pereira, Cristina Bruno (eds) *Catadores de materiais reciclaveis: Um encontro nacional.* Rio de Janeiro: ipea.

Marello, Marta and Ann Helwge. 2014. *Solid Waste Management and Social Inclusion of Waste Pickers: Opportunities and Challenges.* Global Economic Governance Initiative (GEGI) Working Paper, September, Paper 7. Boston, MA: GEGI.

Martín López, Tara. 2014. *The Winter of Discontent: Myth, Memory and History.* Liverpool: Liverpool University Press.

Marx, Karl. 1975 [1852]. *The Eighteenth Brumaire of Louis Bonaparte.* New York: C.P. Dutt.

Marzec, Robert P. 2015. *Militarizing the Environment: Climate Change and the Security State.* Minneapolis, MN: University of Minnesota Press.

Mason, Bob and Peter Bain. 1991. Trade union recruitment strategies: Facing the 1990s. *Industrial Relations Journal* 22(1): 36–45.

Matonte, Cecilia, Lucía Fernández and Martín Sanguinetti. 2017. Entre ideología y gestión. *Brecha*, 17 March. Available at: https://brecha.com.uy/entre-ideologia-y-gestion/

May, Christopher. 2000. *A Global Economy of Intellectual Property Rights: The New Enclosures?* Abingdon: Routledge.

Mayhew, Henry. 1968 [1851]. *London Labour and the London Poor*, vol. II. London: Dover Publications.

Mbembe, Achille. 1992. The banality of power and the aesthetics of vulgarity in the postcolony. *Public Culture* 4(2): 1–30.

Medina, Martín. 2000. Scavenger cooperatives in Asia and Latin America. *Resources, Conservation, and Recycling* 31(1): 51–69.

Medina, Martín. 2005. *The Waste Picker Cooperatives in Developing Countries.* Working Paper. Available at: www.wiego.org/sites/default/files/publications/files/Medina-wastepickers.pdf (accessed 1 April 2020).

Melosi, Martin. 1995. Equity, eco-racism and environmental history. *Environmental History Review* 19(3): 1–16.

Mercosur Press. 2020. Argentina pressing Uruguay on Falklands' RAF flights landing in Carrasco. 21 February. Available at: https://en.mercopress.com/2020/02/21/argentina-pressing-uruguay-on-falklands-raf-flights-landing-in-carrasco (accessed 20 August 2021).

Metcalfe, Alan et al. 2012. Food waste bins: Bridging infrastructures and practices. *Sociological Review* 60: 135–155.

Millar, Kathleen. 2012. Trash ties: urban politics, economic crisis and Rio de Janeiro's garbage dump. In: Katherine Alexander and Joshua Reno (eds) *Economies of Recycling*. London: Zed Books.

Millar, Kathleen. 2014. The precarious present: Wageless labor and disrupted life in Rio de Janeiro, Brazil. *Cultural Anthropology* 29(1): 32–53.

Millar, Kathleen. 2015. The tempo of wage-less work: E.P Thompson's time-sense at the edges of Rio de Janeiro. *Focaal* 73: 28–40.

Millar, Kathleen. 2018. *Reclaiming the Discarded: Life and Labor on Rio's Garbage Dump*. Durham, NC: Duke University Press.

Millar, Kathleen. 2020. Garbage as racialization. *Anthropology and Humanism* 45(1): 4–24.

Millot, Julio and Magdalena Bertino. 1996. *Historia Económica del Uruguay*, tomo II. Montevideo: Fundación de Cultura Universitaria.

Mills, Mary Beth. 2003. Gender and inequality in the global labor force. *Annual Review of Anthropology* 32: 41–62.

Miraftab, Faranak. 2004. Neoliberalism and casualization of public sector services: The case of waste collection services in Cape Town, South Africa. *International Journal of Urban and Regional Research* 28(4): 974–992.

Mohai, Paul, David Pellow and Roberts, J. Timmons. 2009. Environmental justice. *Annual Review of Environment and Resources* 34: 405–430.

Molina, Aurelio. 1980. Un desesperanzado grupo humano agudiza la tristeza del basural. *El Día*, 20 January, p. 18.

Mollona, Mao. 2005. Gifts of labour: Steel production and technological imagination in an area of urban deprivation, Sheffield, UK. *Critique of Anthropology* 25(2): 177–198.

Moore, Sarah A. 2009. The excess of modernity: Garbage politics in Oaxaca, Mexico. *Professional Geographer* 61(4): 426–437.

Moraes, María Inés. 2008. *La Pradera Perdida: Historía y economía del agro Uruguayo – una vision al largo plazo, 1760–1970*. Montevideo: Linardi y Risso.

MTE Argentina. 2013. Marcha MTE y Federación de Cartoneros y Recicladores (CTEP). Available at: www.youtube.com/watch?v=Ckq76rbHGTU (accessed 16 April 2020).

Murray de López, J. 2015. Conflict and reproductive health in urban Chiapas: Disappearing the partera empírica. *Anthropology Matters* 16: 1.

Nandini, Gooptu. 2001. *The Politics of the Urban Poor in Early Twentieth-century India*. Cambridge: Cambridge University Press.

Narotsky, Susana. 2020. *Introduction: Grassroots Economics in Europe*. In: Susana Narotsky (ed.) *Grassroots Economies: Living with Austerity in Southern Europe*. London: Pluto Press.

Narotsky, Susana and Niko Besnier. 2014. Crisis, value, and hope: Rethinking the economy, *Current Anthropology* 55(S9): S4–S16.

Neeson, J.M. 1993. *Commoners: Common Right, Enclosure and Social Change in England, 1700–1820*. Cambridge: Cambridge University Press.

Negrao, Marcelo Pires. 2014. Urban solids waste are commons? A case study in Rio de Janeiro region, Brazil. Paper presented at the Ostrom Workshop (WOW5) conference, Indiana University Bloomington, 18–21 June.

Neiburg, Federico and Natacha Nicaise. 2010. *Garbage, Stigmatization, Commerce, Politics: Port au Prince, Haiti*. Rio de Janeiro: Viva Rio.

Neilson, Brett and Ned Rossiter. 2008. Precarity as a political concept, or Fordism as exception. *Theory, Culture & Society* 25(7–8): 51–72.

Nguyen, Minh T.N. 2018. *Waste and Wealth: An Ethnography of Labor, Value, and Morality in a Vietnamese Recycling Economy*. Oxford: Oxford University Press.

Nixon, Rob. 2013. *Slow Violence and the Environmentalism of the Poor*. Cambridge, MA: Harvard University Press.

Noble, Patricia. 2019. Circular economy and inclusion of informal waste pickers. In: Patrick Schroder, Mnisha Anantharaman, Kartika Anggraeni and Timothy J. Foxon. *The Circular Economy and the Global South: Sustainable Lifestyles and Green Industrial Development*. Abingdon: Routledge.

O'Brien, Martin. 2007. *A Crisis of Waste? Understanding the Rubbish Society*. London: Routledge.

O'Connor, Clementine and Manuela Gheoldus. 2014. *Comparative Study on EU Member States' Legislation and Practices on Food Donations*. Brussels: European Economic and Social Committee/ Bio by Deloitte.

OED (*Oxford English Dictionary*). 2020a. Waste. Available at: www.oed.com/view/Entry/226029?rskey=LKbeD1&result=4#eid (accessed 1 April 2020).

OED (*Oxford English Dictionary*). 2020b. Sedentary. Available at: www.oed.com/view/Entry/174685?redirectedFrom=sedentary#eid (accessed 10 April 2020).

O'Hare, Patrick. 2013. *On the Road to Recovery: Recovery, Transformation and Emergence in an Argentine Recycling Cooperative*. MRes thesis in Social Anthropology, University of Cambridge.

O'Hare, Patrick. 2017. The Uruguayan recyclers' union: Clasificadores, circulation, and the challenge of mobile labor organizations. In: Trevor Ngwane, Immanuel Ness and Luke Sinwell (eds) *Urban Revolt: State Power and the Rise of People's Movements in the Global South*. New York: Haymarket.

O'Hare, Patrick. 2018. 'The landfill has always borne fruit': Precarity, formalisation and dispossession among Uruguay's waste pickers. *Dialectical Anthropology* 43: 31–44.

O'Hare, Patrick. 2020. 'We looked after people better when we were informal': The 'quasi-formalisation' of Montevideo's waste-pickers. *Bulletin of Latin American Research* 39(1): 53–68.

OPP (Oficina de Planeamiento y Presupuesto), Fichtner y LKSur Asociados. 2005. Plan Director de Residuos Sólidos de Montevideo y Área Metropolitana.

Ostrum, Elinor. 1990. *Governing the Commons*. Cambridge: Cambridge University Press.

Ostrum, Elinor and Charlotte Hess. 2007. Private and common property rights. Paper presented at Workshop in Political Theory and Policy Analysis, University of Indiana.

Ostrum, Elinor, Roy Gardener and James Walker. 1994. *Rules, Games and Common-pool Resource*. Ann Arbor, MI: University of Michigan Press.

Packaging Europe. 2020. Reuse: A closer look at Coca-Cola Brazil's unique returnable bottle initiative. 11 February. Available at: https://packagingeurope.com/coca-cola-brazil-returnable-bottle-initiative/ (accessed 1 April 2020).

Palomera, Jamie and Theodora Vetta. 2016. Moral economy: Rethinking a radical concept. *Anthropological Theory* 16(4): 413–432.

Passos Lima, María Raquel. 2015. *O Avesso do Lixo: Materialidade, valor e visibilidade*. PhD thesis, Programa de Pós-Graduación em Sociología e Antropología do Instituto de Filosofía e Ciencias Sociais da Universidade Federal do Rio de Janeiro.

Peattie, Lisa. 1987. An idea in good currency and how it grew: The informal sector. *World Development* 15(7): 851–860.

Pendle, George. 1952. *Uruguay: South America's First Welfare State*. London: Royal Institute of International Affairs.

Peters, Pauline E. 1987. Embedded systems and rooted models: The grazing lands of Botswana and the commons debate. In: Bonnie J. McCay and James M. Acheson (eds) *The Question of the Commons: The Culture and Ecology of Communal Resources*. Tucson: University of Arizona Press.

PNUD (Programa de Naciones Unidas de Desarrollo) and PNUMA (Programa de las Naciones Unidas para el Medio Ambiente). 2012. Implementación de la Ley de Envases: Informe de evaluación. Montevideo: PNUD/PNUMA.

Polanyi, Karl. 2001 [1944]. *The Great Transformation: The Political and Economic Origins of Our Time*. Boston, MA: Beacon Press.

PUC (Plan Uruguay Clasifica). 2006. *Tirando del carro*. Montevideo: MIDES.

PUC (Plan Uruguay Clasifica). 2008. *Clasificar para incluir, incluir para reciclar*. Montevideo: MIDES.

RAE (Real Academia Española). 2020. Trasto. Available at: www.rae.es/drae2001/trasto (accessed 7 April 2020).

Rancière, Jacques. 2004. *The Politics of Aesthetics: The Distribution of the Sensible*. London: Continuum.

Ras, Norberto. 1996. *El gaucho y la ley*. Montevideo: Carlos Marchesi.

Rathje, William and Cullen Murphy. 2001. *Rubbish: The Archaeology of Garbage*. Tucson: University of Arizona Press.

Red-LACRE (Red Latinoamericano de Recicladores). 2017. *Análisis de políticas públicas para el reciclaje inclusive en América Latina*. Febrero. Available at: www.rds.org.co/es/novedades/analisis-de-politicas-publicas-para-el-reciclaje-inclusivo-en-america-latina (accessed 20 August 2021).

Renfrew, Daniel. 2009. In the margins of contamination: Lead poisoning and the production of neoliberal nature in Uruguay. *Journal of Political Ecology* 16(1): 87–103.

Reno, Joshua O. 2009. Your trash is someone's treasure. *Journal of Material Culture* 14(1): 29–46.

Reno, Joshua O. 2014. Toward a new theory of waste: From 'matter out of place' to signs of life. *Theory, Culture & Society* 31(6): 3–27.

Reno, Joshua O. 2015. Waste and waste management. *Annual Review of Anthropology*. 44: 557–572.

Reno, Joshua O. 2016. *Waste Away: Working and Living with a North American Landfill*. Berkeley: University of California Press.

Robbins, Bruce. 2007. The smell of infrastructure. *Boundary* 34(1): 25–33.

Robbins, Joel. 2013. Beyond the suffering subject: Toward an anthropology of the good. *Journal of the Royal Anthropological Institute* 19: 447–462.

Robinson, Cedric J. 2000 [1983]. *Black Marxism: the Making of the Black Radical Tradition*. Chapel Hill: University of North Carolina Press.

Rodgers, Christopher. 2016. *Contested Common Land: Environmental Governance Past and Present*. London: Earthscan.

Rogaski, Ruth. 2004. *Hygienic Modernity: Meanings of Health and Disease in Treaty-Port China*. Berkeley: University of California Press.

Roman, Carolina. 2016. Evolución del consume privado histórico: 1870–1955. Paper presented at the VI Jornadas Académicas de la Facultad de Ciencias Económicas y de Administración, Universidad de la República, Montevideo.

Rosaldo, Manuel. 2016. Revolution in the garbage dump: The political and economic foundations of the Colombian recycler movement, 1986–2011. *Social Problems*63(3): 351–372.

Samson, Melanie. 2015a. *Forging a New Conceptualization of 'the Public' in Waste Management*. WIEGO Working Paper No. 32. Manchester: WIEGO.

Samson, Melanie. 2015b. Accumulation by dispossession and informal economy: Struggles over knowledge, being and waste at a Soweto garbage dump. *Environment and Planning D: Society and Space* 33(5): 813–830.

Sanchez, Andrew. 2012. Deadwood and paternalism: Rationalising casual labour in an Indian company town. *Journal of the Royal Anthropological Institute* 18(4): 808–827.

Sarachu, Gerardo and Fernando Texeira. 2013. ¿Escribanos del deterioro? Reflexiones sobre los límites de la intervención universitaria junto a colectivos de trabajadores y trabajadoras de la clasificación de residuos en Montevideo. *Revista Estudios Cooperativos* 18: 111–132.

Sargent, Frederic O. 1958. The persistence of communal tenure in French agriculture. *Agricultural History*, 32(2): 100–108.

Scanlan, John. 2005. *On Garbage*. London: Reaktion.

Schamber, Pablo. 2008. *De los deshechos a las mercancías: Una etnografía de los cartoneros*. Buenos Aires: Paidós.

Schober, Elizabeth. 2016. *Base Encounters: The US Armed Forces in South Korea*. London: Pluto Press.

Schulz, Yvan and Anna Lora-Wainwright. 2019. In the name of circularity: Environmental improvement and business slowdown in a Chinese recycling hub. *Worldwide Waste* 2(1): 1–13.

Schumpeter, Joseph A. 1994 [1942]. *Capitalism, Socialism and Democracy*. London: Routledge.

Scott, James C. 1985. *Weapons of the Weak: Everyday Forms of Peasant Resistance*. New Haven, CT: Yale University Press.

Shever, Elana. 2012. *Resources for Reform: Oil and Neoliberalism in Argentina*. Stanford, CA: Stanford University Press.

Silverman, Jana. 2011. Labor relations in Uruguay under the Frente Amplio government, 2005–2009: From neoliberalism to neocorporativism? Paper presented at VII Global Labour University Conference, University of Witwatersrand, South Africa.

Simone, AbdouMaliq. 2004. People as infrastructure: Intersecting fragments in Johannesburg. *Public Culture* 16(3): 407–429.

Simpson, Bob. 1994. Bringing the unclear family into focus: Divorce and re-marriage in contemporary Britain. *Man* 29(4): 831–851.

Skelton, Leona Jayne. 2012. *Environmental Regulation in Edinburgh and York; c.1560–c.1700 with Reference to Several Scottish Burghs and Northern English Towns*. PhD thesis, Durham University. Available at Durham E-Theses Online: http://etheses.dur.ac.uk/7016/

Smith, Adam. 1896. *Lectures on Justice, Police, Revenue and Arms (1763)*. Oxford: Clarendon Press.

Solari, Aldo E. 1953. *Sociología rural nacional*. Montevideo: M.B. Altura.

Sorroche, Santiago. 2015. *Gubernmentalidad global y vernacularización en la gestión de residuos. Análisis etnográfico desde la experiencia de cooperativas de cartoneros en la Gran Buenos Aires*. Doctoral thesis, Universidad de Buenos Aires.

Stallabrass, Peter. 1990. Marx and heterogeneity: Thinking the lumpenproletariat. *Representations* 31: 69–95.

Stamatopoulou-Robbins, Sophia. 2014. Occupational hazards. *Comparative Studies of South Asia, Africa, and the Middle East* 43(3): 476–496.

Stamatopoulou-Robbins, Sophia. 2020. *Waste Siege: The Life of Infrastructure in Palestine*. Stanford, CA: Stanford University Press.

Standing, Guy. 2011. *The Precariat: The New Dangerous Class*. London: Bloomsbury Academic.

Sternberg, Ana Carolina. 2013. From 'cartoneros' to 'recolectores urbanos': The changing rhetoric and urban waste management policies in neoliberal Buenos Aires. *Geoforum* 48: 187–195.

Strasser, Susan. 1999. *Waste and Want*. New York: Metropolitan Books.

Strathern, Marilyn. 1991. *Partial Connections*. Lanham, MD: Rowman and Littlefield.

Suárez, Francisco M. 2016. *La Reina del Plata, Buenos Aires: Sociedad y residuos*. Buenos Aires: Ediciones UNGS.

Suárez, Francisco M. et al. 2014. *Fisuras: Dos estudios sobre pasta base de cocaína en el Uruguay*. Montevideo: Universidad de la República.

Susser, Ida and Stéphane Tonnelat. 2013. Transformative cities. *Focaal* 66: 105–121.

Svampa, Maristella. 2004. La experiencia piquetera: El desafío de las organizaciones de desocupados en Argentina. *Revista da Sociedade Brasileira da Economía Política* 15: 88–110.

Telemundo. 2016. Ex director de Limpieza: El problema central de Montevideo es la clasificación ilegal de residuos. 5 May. Available at: www.teledoce.com/telemundo/nacionales/ex-director-de-limpieza-el-problema-central-de-montevideo-es-la-clasificacion-ilegal-de-residuos/ (accessed 1 April 2020).

Tenner, Edward. 1996. *Why Things Bite Back: Technology and the Revenge of Unintended Consequences*. New York: Vintage.

Thoburn, Nicholas. 2003. *Deleuze, Marx, Politics*. London: Routledge.

Thompson, E.P. 1991. *Customs in Common*. London: Penguin.

Thompson, Michael. 2017 [1979]. *Rubbish Theory*. Oxford: Oxford University Press.

Thompson, Vivienne E. 2009. *Garbage In, Garbage Out: Solving the Problems with Long-distance Trash Transport*. Charlottesville: University of Virginia Press.

Thurgood, Maggie. 1999. *Solid Waste Landfills: Decision-makers' Guide Summary*. Washington/Copenhagen: World Bank, WHO, SDC, SKAT joint publication.

Tisdale, Richard. 2018. 1948. Stinking sewage scavengers discovered. Available at: https://newsfromthepastblog.wordpress.com/2018/11/27/1848-stinking-sewage-scavengers-discovered/ (accessed 20 August 2021).

Tsing, Anna. 2015. *The Mushroom at the End of the World: On the Possibility of Life in Capitalist Ruins*. Princeton, NJ: Princeton University Press.

Última Hora. 1973. Y la basura se sigue amontonando. 4 January.

Últimas Noticias. 1986. A tres cuadros del centro y a siglos de distancia. 3 April.

UPI (United Press International). 2011. Uruguay's solidarity with Argentina over Falklands comes with costs, 20 December. Available at: www.upi.com/Energy-News/2011/12/20/Uruguays-solidarity-with-Argentina-over-Falklands-comes-with-costs/90221324420486/ (accessed 20 August 2021).

Van Loon, Joost. 2002. *Risk and Technological Culture: Towards a Sociology of Virulence*. London: Routledge.

Vargas-Cetina, Gabriela. 2005. Anthropology and cooperatives: From the community paradigm to the ephemeral association in Chiapas, Mexico. *Critique of Anthropology*, 25(3): 229–251.

Vea. 1971. MOP: Obras por 1.800 millones. 12 January.

Venkatesan, Soumhya, Laura Bear, Penny Harvey, Sian Lazar, Laura M. Rival and AbdouMaliq Simone. 2016. Attention to infrastructure offers a welcome reconfiguration of anthropological approaches to the political. *Critique of Anthropology* 38(1): 3–52.

Veolia. 2017. In Mexico City, Veolia will build and operate one of the largest waste-to-energy facilities in the world and the first in Latin America. 22 May

17. Available at: www.veolia.com/en/news/waste-to-energy-renewable-energy-mexico (accessed 1 April 2020).
Viglietti, Daniel. 1968. A desalambrar. In: Daniel Viglietti, *Canciones para el hombre nuevo*. Montevideo: Orfeo.
Vitale, Patrick. 2011. Wages of war: Manufacturing nationalism during World War II. *Antipode* 43(3): 793–819.
Von Schnitzler, Antina. 2013. Travelling technologies: Infrastructure, ethical regimes, and the materiality of politics in South Africa. *Cultural Anthropology* 28(4): 670–693.
Walter, Nicholas, Phillipe Bourgois and H. Margarita Loinaz. 2004. Masculinity and undocumented labor migration: Injured day laborers in San Francisco. *Social Science & Medicine* 59(6): 1159–1168.
Whitson, Risa. 2011. Negotiating place and value: Geographies of waste and scavenging in Buenos Aires. *Antipode* 43(4): 1404–1433.
Williams, Raymond. 1977. *Marxism and Literature*. Oxford: Oxford University Press.
Wirth, Lolí Gewehr. 2016. *Movimento de catadores e a política nacional de resíduos sólidos: A experiencia de Rio Grande do Sul*. PhD thesis, Universidade Estadual de Campinas, Instituto de Filosofía e Ciencias Humanas, Campinas, São Paulo State.
Wolf, Eric. 1971. Introduction. In: Norman Miller and Roderick Aya (eds) *National Liberation: Revolution in the Third World*. New York: The Free Press.
Wolman, Abel. 1965. The metabolism of cities. *Scientific American* 213(3): 179–190.
World Economic Forum, Ellen MacArthur Foundation and McKinsey & Company. 2016. *The New Plastics Economy – Rethinking the Future of Plastics*. Available at: www.ellenmacarthurfoundation.org/publications
Worsley, Peter. 1971. Introduction. In: Peter Worsley (ed.) *Two Blades of Grass: Rural Cooperatives in Agricultural Modernization*. Manchester: Manchester University Press.
Yaeger, Patricia. 2003. Trash as archive: Trash as enlightenment. In: Gay Hawkins and Stephen Muecke (eds) *Culture and Waste: The Creation and Destruction of Value*. Lanham, MD: Rowman and Littlefield.
Yaffé, Jaime. 2009. La economía de la dictadura. Paper presented at the VIII Congreso Brasileiro de Historia Económica (Campinas, 8–9 September 2009).
Yanagisako, Sylvia. 2002. *Producing Culture and Capital: Family Firms in Italy*. Princeton, NJ: Princeton University Press.
Ye, Junjia. 2014. Migrant masculinities: Bangladeshi men in Singapore's labour force. *Gender, Place, and Culture* 21(8): 1012–1028.
Yeo, Eileen. 1971. Mayhew as a social investigator. In: E.P. Thompson and Eileen Yeo, *The Unknown Mayhew: Selections from the Morning Chronicle, 1849–1850*. London: Merlin Press.
Zapata, Patrik and María José Zapata Campos. 2015. Producing, appropriating and recreating the myth of the urban commons. In: Christian Borch and Martin Kornberger (eds) *Urban Commons: Rethinking the City*. London Routledge.

Index

ill refers to an illustration

PGIL2021USA